OXFORD SURVEYS IN
EVOLUTIONARY BIOLOGY

Volume 6
1989

OXFORD SURVEYS IN EVOLUTIONARY BIOLOGY

EDITED BY
PAUL H. HARVEY AND LINDA PARTRIDGE

Volume 6
1989

OXFORD UNIVERSITY PRESS
1989

Oxford University Press, Walton Street, Oxford OX2 6DP
Oxford New York Toronto
Delhi Bombay Calcutta Madras Karachi
Petaling Jaya Singapore Hong Kong Tokyo
Nairobi Dar es Salaam Cape Town
Melbourne Auckland
and associated companies in
Berlin Ibadan

Oxford is a trade mark of Oxford University Press

Published in the United States
by Oxford University Press, New York

© Oxford University Press, 1989

British Library Cataloguing in Publication Data
Oxford surveys in evolutionary biology
Vol. 6 1989
1. Organisms. Evolution—Serials
575'.005
ISBN 0–19–854252–6

Library of Congress Cataloging in Publication Data

(LC Card number 86–641417)

Typeset by Latimer Trend & Company Ltd, Plymouth
Printed in Great Britain by Courier International, Tiptree

Contents

Contributors

AUSTIN BURT, Department of Biology, McGill University, 1205 Avenue Docteur Penfield, Montreal, P.Q., H3A 1B1, Canada

DEBORAH M. GORDON, Department of Zoology, University of Oxford, South Parks Road, Oxford OX1 3PS, UK

PAUL H. HARVEY, Department of Zoology, University of Oxford, South Parks Road, Oxford OX1 3PS, UK

ANDREW J. LEIGH BROWN, Department of Genetics, University of Edinburgh, West Mains Road, Edinburgh EH9 3JN, Scotland, UK

JOHN MAYNARD SMITH, School of Biological Sciences, University of Sussex, Falmer, Brighton, Sussex BN1 9QG, UK

DANIEL E. L. PROMISLOW, Department of Zoology, University of Oxford, South Parks Road, Oxford OX1 3PS, UK

DAVID C. QUELLER, Department of Ecology and Evolutionary Biology, Rice University, P.O. Box 1892, Houston, Texas 77251, USA

ANDREW F. READ, Department of Zoology, University of Oxford, South Parks Road, Oxford OX1 3PS, UK

EÖRS SZATHMÁRY, Ecological Modelling Research Group, Department of Plant Taxonomy and Ecology, Roland Eötvös University, 1083, Budapest, Kun Bela ter., 2, Hungary

ELISABETH S. VRBA, Department of Geology and Geophysics, Yale University, Kline Geology Laboratory, P.O. Box 6666, CT 06511-8130, USA

Weismann and modern biology

JOHN MAYNARD SMITH

1. INTRODUCTION

In the preface of *Back to Methuselah*, Shaw describes how Weismann cut the tails off mice, in order to demonstrate the non-inheritance of acquired characters. Not only cruel, Shaw exclaims, but stupid. Weismann should have known that dog fanciers have been docking the tails of bitches for generations without the smallest effect on their offspring. In any case, Lamarckists would expect only actively acquired changes to be inherited: the surgical loss of a tail is not such a change. As it happens, Shaw is being unfair to Weismann's brains, if not to his heart. Weismann describes how, when he first put forward the view that acquired characteristics are not inherited, he was met by a chorus of objections from critics who pointed to the known inheritance of mutilations, including the birth of tailless puppies to mothers whose tails had been docked. He adds 'even student's fencing scars were said to have been occasionally transmitted to their sons (happily not to their daughters)' (2,65: this and other volume and page references are to Weismann 1904). In refuting this criticism, Weismann referred to the absence of documented evidence for the effects claimed, and to the ineffectiveness of such human practices as circumcision; he also performed his famous mouse experiment.

Nevertheless, when I read Shaw as a boy, it left me with the impression that Weismann was a cruel and ignorant German pedant, an impression that was slow to fade. Until relatively recently, I knew, or thought I knew, only three things about Weismann's contribution to science. First, he had argued for the non-inheritance of acquired characters on the grounds of early segregation of the germ line from the soma—an argument that carries little weight, as there is no segregated germ line in plants, yet Lamarckian inheritance is as rare in plants as in animals. Secondly he had explained the evolution of ageing on the manifestly fallacious grounds that, if individuals did not die, there would be no room for the operation of natural selection. Thirdly, he had proposed an erroneous theory of development, according to which the cells in a particular organ—the liver, say—contain only those genes relevant for liver function.

This chapter is an attempt to make amends. I have gradually become aware, that, after Darwin, Weismann was the greatest evolutionary biologist of the nineteenth century. Further, the problems he was concerned with are often the same problems that concern us today. I have neither the linguistic nor the historical skills to write a proper history of his ideas. Instead, I have attempted to summarize what he thought towards the end of his life, when he wrote *The evolution theory*, and to explain why he thought as he did. In the light of present knowledge, it is sometimes easy to see where he went wrong; it is harder to see what he got right, because when he did so, his ideas are now so widely accepted that we do not appreciate that people once thought otherwise. For this reason, the following account may appear unduly critical. It is not intended to be. Indeed, the main impression I am left with is of a man who spent his life thinking about precisely the problems that have concerned me during my own work, who often got it exactly right, and who was admirably willing to admit that he had been wrong and to try again.

I cannot, in a short chapter, discuss all Weismann's ideas. I have therefore omitted any mention of his discussion of the nature and origin of species, partly because this is not a topic to which he made any major contribution, and partly because his ideas have recently been discussed by Mayr (1988). I also say little of his work on the nature of adaptation, particularly in the Lepidoptera. This is a serious distortion. Weismann, just as Darwin, was a naturalist of wide knowledge, and a man in love with nature. As such naturalists tend to be, he was deeply impressed by the ubiquity and complexity of adaptation. The opening chapters of his book deal with such topics as animal coloration, mimicry, carnivorous plants, animal instincts, symbiosis, and the function of flowers. He also treats sexual selection, accepting Darwin's argument in its essentials. The point of these chapters is to bring home to the reader that the phenomenon that any theory of evolution must explain is that of adaptation. I shall not discuss these chapters any further, admirable as they are, because his opinion does not differ essentially from that of Darwin and Wallace. What is new in Weismann is a more sophisticated theory of development and heredity, and it is on this that I shall concentrate.

2. THE GERM PLASM THEORY

In formulating his theory of inheritance, Weismann had the advantage over Darwin of important discoveries in cytology. He knew that, in mitosis, chromosomes split into two apparently equal parts, and that (in most cases) the daughter cells contain the same number (and, he assumed, kinds) of chromosomes as the parent cell: that, at fertilization, the nuclei of two gametes, each containing half the typical chromosome number, fuse to form the new zygotic nucleus: and, his own contribution, that in the meiotic divisions, the chromosome number is halved. He assumed, as did many others, that the chromosomes were the carriers of the hereditary material.

In describing his 'germ plasm' theory, he uses some technical terms. A 'biophor' is the smallest unit of replication: it is 'the living molecule ... which possesses the marvellous property of growth and division into two halves similar to itself and to the ancestral molecule' (1,275). In contrasting a 'living' molecule with 'an ordinary chemical molecule', the essential difference is that the former can replicate. A 'determinant' is a group of biophors which can cause the appearance of particular properties in particular kinds of cell. It is important for him that the biophors do not contain the actual properties (e.g. in modern terms, he does not imagine that a determinant of a muscle cell contains muscle proteins, but only the capacity to cause the appearance of muscle protein). Finally, an 'id' is a complete set of determinants—in modern terms, a haploid genome, although, as we shall see, Weismann had no concept of haploidy. He insists that determinants are qualitatively distinct and that a cell lacking a determinant cannot acquire it *de novo*. Therefore a gamete must contain at least one complete id.

Allowing for the unfamiliarity of his terminology and his ignorance of molecular biology, all this is very modern—in particular, his insistence that a 'living' entity must have the properties of qualitative distinctness, and ability to replicate. He then turns to the problem of development. He describes the old argument between preformation and epigenesis, and admits that this has been settled in favour of epigenesis: there is no homunculus in the sperm. Yet he says that his own ideas are a return to the concept of preformation, but in a new form: the zygote may not contain a homunculus, but it does contain a set of determinants that specify how development shall proceed. In defending this position, he stresses that any theory of development must also explain inheritance. He was much struck by the inheritance of localized characteristics. For example, he mentions that the butterfly, *Lycaena agestis*, has a black spot on the wing in Europe, but a white spot in Scotland (1,356: it is characteristic that he defends his theoretical position by reference to facts about butterflies). He concludes that the determinants 'each stand in a definite relationship to particular cells or kinds of cells'. He imagines that this is brought about by unequal division of the genetic material during development, so that each determinant 'must be guided through the numerous cell divisions of ontogeny, so that it shall ultimately come to lie in the cells which it is to control' (1,373).

Today we would see this as an error. Instead, we hold that, as a rule, all genes are transmitted to all cells, but that different genes are activated in different cells. Weismann was well aware of this alternative. He writes 'From the very first, therefore, I have considered whether it would not be better to elaborate the determinant theory in such a way that it would not be necessary to assume a disintegration of the id [i.e. unequal division of the genetic material, J.M.S] in the course of ontogeny, but simply to conceive of every expression of activity on the part of a determinant as dependent on a specific stimulus ... I rejected this hypothesis because of the enormous number of specific stimuli which it demands' (1,382). Thus Weismann was well aware of our view as a conceivable alternative. If he rejected it, he did so for good reason. It is a striking fact that we are still quite unable to account for the 'enormous number of specific stimuli', to such an extent that many biologists seem unaware that the problem exists. Interestingly, Weismann did not regard his choice on this issue—differential transmission of genes rather than differential activation—as fundamental. He writes, 'As to the fundamental ideas expressed in the theory, I have already shown that these remain unaltered, even if we do not assume a disintegration or segregation of the germ-plasm, but think of all developing cells as equipped with the complete germ-plasm' (1,407). What was essential was the idea of replicating determinants influencing particular cell types during ontogeny.

Difficulties arose for his theory from two sets of facts. First, experiments showed that isolated blastomeres can sometimes give rise to a complete embryo. He recognized the distinction between mosaic and regulative development, and explained the latter by assuming a later segregation of the determinants. The facts of regeneration posed a more serious threat: if a part can give rise to a whole, then the part must have had a complete genome. For example, he recognizes that any portion of plant cambium can give rise to a whole plant, and concludes that cambium cells must contain an 'accessory idioplasm'. Reading his discussion of regeneration, one gets the impression of a theory in crisis.

The evidence concerning the existence of a 'germ track', from which alone new germ cells could arise, was important to him for two reasons. First, it was a 'corroboration' of his view that 'germ-plasm cannot be produced *de novo*' (1,411). If somatic cells contain only an incomplete genome, they cannot give rise to gametes. Secondly, it provided an explanation of 'natural death'—of what we would call senescence. Somatic cells can have no long-term future, as they lack a complete genome. He argues, therefore, that '. . . every function and every organ disappears as soon as it becomes superfluous . . . the power of being able to live on without limit is useless for the somatic cells, and thus also for the body, since they cannot produce new reproductive cells after those that are present are liberated: and with this the individual ceases to be of any value for the preservation of the species' (1,261). This is not a fully satisfactory explanation of senescence: why not an immortal soma, carrying an indefinitely reproducing germ line? However, his views on senescence are not fully explained in *The evolution theory*. Fortunately, they have been discussed at some length by Kirkwood and Cremer (1982), whose account I now summarize.

In Weismann's first attempt at a theory of ageing, published in 1882, he argued that individuals would inevitably suffer injuries which could not be repaired, and it would therefore be to the benefit of the species that such individuals should senesce and die. This is, in Kirkwood and Cremer's terminology, an 'adaptive' theory of senescence. That is, it ascribed a positive selective advantage to the process of senescence *itself*: it does not explain senescence as the unselected consequence of other changes that are themselves selected. It is also a circular argument, as was forcefully pointed out by Medawar (1952): it explains decay by assuming that old animals are decayed. However, in his later writings, Weismann gave less and less emphasis to this 'adaptive' explanation. In *The evolution theory*, he writes 'I did not at once discover the true explanation (of ageing)', although, unfortunately, he does not tell us why he thinks that his earlier explanation had been mistaken. It is also clear, from the quotation in the last paragraph, that he did not achieve a fully consistent and logical theory of ageing. However he did, in his later writings, propose a theory of ageing very similar to the pleiotropic theory of ageing of Williams (1957), which most of us today would accept. He suggests that somatic cells which become more efficient in the short term in performing their specific tasks may pay the price of losing their potential immortality: 'Who can point out with any feeling of confidence the direct advantage in which somatic cells, capable of limited duration, excelled those capable of eternal duration? Perhaps it was in a better performance of their special physiological tasks, perhaps in additional material and energy available for the reproductive cells as a result of this renunciation of the somatic cells, or perhaps such additional power conferred upon the whole organism a greater power of resistance in the struggle for existence, than it would have had, if it had been necessary to regulate all the cells to a corresponding duration' (Weismann 1891, p. 142). On his best days, then, Weismann had a fully modern theory of the evolution of senescence, but he did not keep to it consistently.

The most serious weakness of his germ plasm theory lies in his ignorance of the fact that, typically, each determinant is represented once, and once only, in a gamete, and that determinants are linked, so that when one is replicated, all are replicated. He thought that each chromosome contained a complete genome, and sometimes several genomes. In one of the very few diagrams in the book (1,348), he shows the consequences of meiosis and syngamy, assuming many genomes (ids) per gamete—in fact 8 per gamete, and 16 per zygote. Ironically, if he had made the logically simplest assumption of 1 id per gamete, he might have rediscovered Mendel's laws *a priori*. As it is, his ideas about genetics, particularly of populations, are needlessly complicated. More serious, as we shall see below, his ignorance of linked replication may have contributed to his misleading views about 'germinal selection'.

3. ARE FUNCTIONAL MODIFICATIONS TRANSMITTED?

Weismann was well aware that functional modifications—for example in the growth of bone, muscle, and tendon—can occur during ontogeny. He explained these modifications in terms of Roux's idea that 'an organ increases through its own specific activity'. He thought that this came about by a kind of selection operating between the cells of an individual, each cell type being stimulated to multiply by appropriate conditions. But he is clear that the properties of the cells that cause them to react appropriately 'are themselves adaptations of the organism, and can therefore be referred to personal selection' (1,248), and 'not from a struggle between the cells themselves' (1,250). He says (1,248) that 'personal selection and histonal selection cooperate', a remark reminiscent of Buss' (1987) argument that evolutionary novelty requires synergy between selection at the cell and organism levels.

But although Weismann accepted the reality of such functional modifications, he was adamant that they were not transmitted via the germ cells. He defends this position by four arguments:

1. There is no empirical evidence for the inheritance of acquired characters.

2. Many adaptations are of a kind that could not arise as functional modifications during ontogeny.

3. Adaptations of the sterile castes of insects could not evolve by a Lamarckian mechanism.

4. There is a theoretical difficulty about Lamarckian inheritance, which we would today describe as the problem of reverse translation.

The first point was discussed briefly above, and the third is persuasive and needs no elaboration. Point 2 is also persuasive. If an organ is fully formed before it is used, and cannot then be modified, there will be no functional modifications to be inherited. Examples of such structures are the colour of many animals (in particular, of his beloved Lepidoptera), the cuticle of arthropods, and the protective thorns and hairs of plants. Of particular interest are instincts, as the idea that instincts might be inherited habits has often been seen as favouring Lamarckism. Weismann points out that there are many instincts that could not have arisen as a learnt habit in the first place, for several reasons. First, the animal in question may be of too low intelligence to learn a complex habit: for example, it is inconceivable that a solitary wasp should learn by trial and error its complex sequence of nest-building and provisioning behaviour. Secondly, many instinctive acts are performed only once in a lifetime, and so afford no opportunity for learning: for example, the acts leading to the suspension of some lepidopteran pupae. Finally, the success or failure of some instinctive acts is decided only after the performer is dead, and so could not guide learning; for example, oviposition in many insects.

If, then, many complex structures and behaviours have evolved without the transmission of functional modifications, and if there is no evidence for such transmission in any case, there is no need to assume that the Lamarckian mechanism has ever been important. But Weismann has a final theoretical argument: in fact, although I have listed it above as a final argument, I have little doubt that, in the genesis of Weismann's own ideas, it came first, and the empirical evidence supporting it followed. The argument can best be put in Weismann's own words: oddly, both quotations use the analogy of translation into Chinese.

'But, as these primary constituents [i.e. the genetic determinants, J.M.S.] are quite different from the parts themselves [i.e. the adult organs, J.M.S], they would require to vary in quite a different way from that in which the finished parts had varied: which is very like supposing that an English telegram to China is there received in the Chinese language' (2,63).

And, discussing whether instincts are inherited habits: 'How could it happen that the constant exercise of memory throughout a lifetime . . . could influence the germ cells in such a way that in the offspring the same brain cells which preside over memory will likewise be more highly developed? . . . if we take our stand upon the theory of determinants, it would be necessary to a transmission of acquired strength of memory that the states of these brain cells should be communicated by the telegraphic path of the nerve cells to the germ cells, and should there modify only the determinants of the brain cells, and should do so in such a way that, in the subsequent development of an embryo from the germ cell, the corresponding brain cells should turn out to be capable of increased functional activity . . . I can only compare the assumption of the transmission of the results of memory-exercise to the telegraphing of a poem, which is handed in in German, but at the place of arrival appears on the paper translated into Chinese' (2,107).

It is hard to imagine a clearer expression of the theoretical difficulty of Lamarckian inheritance. In particular, the use of the information analogy is clear and modern. Not surprisingly, since information-transducing machines were rare in 1900, and as there was no scientific definition of information, Weismann did not maintain a consistently informational concept of the gene, a failure that cost him dear when he came to formulate his ideas of germinal selection and the origin of new variation. Despite the persuasiveness of the theoretical argument, Weismann did not regard it as decisive. There is so much in biology that we do not understand, he says, that we cannot rule out a process merely because we cannot imagine how it could happen: we need empirical arguments as well. He would have agreed with Jacob's (1982) remark about reverse translation: 'Not that such a mechanism is theoretically impossible—simply it does not exist'.

To me, the most surprising thing about the chapters dealing with the transmission of functional modification is the argument that is not there— the dog that did not bark in the night. There is no mention, in this context, of the segregation of the germ track. I do not know how far Weismann's research on the origin of germ cells influenced him in reaching his conclusion (first published in 1883) that acquired characters are not inherited, but it is clear that the argument played little part in his final conclusions. Nor do I see how it could have done. Darwin had proposed his pangenesis theory to account for the supposed 'effect of use and disuse', according to which 'gemmules' carry information from the soma to the germ cells. Weismann starts his discussion of Lamarckism by outlining this theory. He rejects it on theoretical grounds (the first of the two Chinese language quotations immediately follows his discussion of Darwin's theory): the segregation of the germ line would be wholly irrelevant.

4. GERMINAL SELECTION

Weismann was aware that evolution requires a source of variation. He sees that meiotic segregation in a sexual species will generate variability, but that there must also be some source of variation in the determinants themselves. In this he is entirely modern, but his theory of the origin of genetic change ('mutation' had in 1904 a narrower meaning) is that it arises from selection between the determinants within an individual. The idea originated in his struggle to understand the gradual disappearance of rudimentary organs. This phenomenon was seen as a more important one then than now, mainly because it seemed to support the idea of Lamarckian inheritance: organs that are not used grow less during ontogeny, and this explains their disappearance in evolution. Weismann was naturally reluctant to accept this explanation, so he had to find an alternative. He did not think that direct personal selection for reduced size on the grounds of economy could explain it, because 'the variations are too slight to have selective value' (2,118). In fact, he repeatedly argues for a threshold below which natural selection was ineffective: I do not understand why he thought so. He is clear that the maintenance of organismic adaptation requires normalizing selection. Could the absence of normalizing selection on rudimentary organs explain their disappearance, as suggested by Romanes? He does not think so. He agrees that functional adaptation would disappear, but he does not see why there should be an apparently monotonic reduction in size, as new variation would be as likely to increase as to decrease size.

He therefore proposes that there is selection between determinants, in the quantity of 'nutrient' they acquire. A determinant may change so as to increase or decrease the nutrient it acquires (I am not clear whether he thinks of the 'increase' as an increase in what we would call copy number, or merely in effectiveness). He assumes that a determinant that increases within the genome will also cause an increase in the size of the relevant organ. A determinant of a rudimentary organ may increase or decrease, but if it increases it will do so at the expense of other determinants, which influence organs that are still needed. Therefore, increases in the determinants of rudimentary organs will be selected against at the personal level, because of their effects in reducing other determinants.

As an *ad hoc* account of the disappearance of rudimentary organs, the theory has something to recommend it. Unfortunately, Weismann developed it into a general theory of the origin of new genetic variation. He argues that a determinant that has increased will have a tendency to go on increasing (a kind of 'molecular drive'), until the effects at the organism level pass the selective threshold, and personal selection steps in. Contrasting his views with those of Darwin, he writes (2,232) 'The essential differences between Darwin's view ... and my own lies in the fact that Darwin conceived of natural selection working only with variations which are not only due to chance themselves, but the intensification of which also depends in its turn solely upon natural selection, while, according to my view, natural selection works with variational tendencies which become intensified through internal causes'. Nevertheless, he insists that personal selection will always win, but this is because what one observes is individual adaptation. In fact, if there was selection between genes within organisms of the kind he imagines, it is doubtful whether individual adaptation could evolve. The linked reproduction of genes was a precondition for further evolution of organisms.

In his introduction to *The evolution theory*, Weismann writes 'This extension of the principle of selection to all grades of vital units is the characteristic feature of my theories: it is to this idea that these lectures lead, and it is this—in my own opinion—which gives the book its importance. This idea will endure even if everything else in the book should prove transient'. He is surely right that the concept that we would now call 'levels of selection' will endure. Its importance in evolution theory is increasing as our knowledge of the range and variety of replicating entities increases. However, the particular form of his germinal selection theory was unfortunate. He pictures the origin of all genetic variation (what we would now call mutation) as essentially quantitative (+ and − only, in his terms), and established in the first instance by germinal selection, and only later exposed to personal selection. It is a pity that he did not pursue the information analogy he had used in discussing Lamarckism, and picture mutation as copying error. That he did not do so is partly due to his need for a theory that would explain the monotonic reduction in the size of rudimentary organs, and partly (Mayr 1988) because of a philosophically motivated preference for a deterministic theory of mutation.

5. SEX AND BREEDING SYSTEMS

It is not to be expected that Weismann should have had a full understanding of the evolution of sex, as we still are feeling our way to such an understanding, but there is much that he does understand. He is clear that sex is not needed for reproduction, and that, at the cellular level, it is the opposite of reproduction. He understands that meiotic segregation generates variability. He suggests that sex is needed for 'coadaptation'—that is, for the bringing together in a single individual of 'harmonious' variations that originated in different individuals. In effect, this is the idea proposed by Fisher (1930) and Muller (1932), that sex accelerates evolution by bringing together, by genetic recombination, favourable mutations from different ancestors. He recognizes that this is a long-term rather than a short-term explanation (2,198), and (2,199) suggests a more immediate advantage in terms of normalizing selection, which, he suggests, can only act effectively to maintain adaptation in a sexual population. I am unable to follow his argument on this point—it is not one that can adequately be presented in verbal terms—and I suspect that it is fallacious. It is true that, for different reasons, Muller (1964) and Kondrashov (1982) have argued that recombination can reduce the genetic load needed to eliminate harmful mutation, but I do not think that Weismann can be credited with reaching this conclusion. However, it is interesting that he does discuss the advantages of sex in the two contexts of directional selection, and the elimination of harmful variation.

He argues that the origin of sex requires a 'direct' (i.e. immediate) selective advantage. He rejects the idea the sex has a rejuvenating function, because of the survival of parthenogens (he cultured a parthenogenetic Ostracod for 80 generations). He suggests that the selective advantage responsible for the origin of cellular fusion may have been that each partner contributed genetic material deficient in the other: an explanation in terms of complementation that most of us today would accept. He is aware of inbreeding depression, and explains it by saying that, in an inbred population, 'the germ plasm may then consist entirely of identical ids' (2,273). He contrasts this with the absence of deterioration in parthenogens which lack a reduction division. He argues that hermaphroditism is favoured in sedentary organisms because of the difficulty gonochorists would experience in finding mates. Finally, and for me annoyingly, he writes 'By the occurrence of parthenogenesis, the number of ova produced by a particular colony of animals may be doubled, because each individual is a female' (2,243). This is a fairly impressive list of insights for a book published in 1904.

6. WEISMANN AND THEORETICAL BIOLOGY

Darwin's theory of evolution by natural selection is the central theory of biology. Yet Darwin was a shame-faced theorist. He found it necessary to pretend to others, and even to himself, that he reached his conclusions by induction from a vast array of facts. Weismann, like Darwin, was first and foremost a naturalist, but he is a less ashamed theorist. In a sense, he is the first conscious theoretical biologist. His failing eyesight may have contributed to this tendency. However, he is a theoretician without the tools of a theoretician's trade. There is, I think, not one line of algebra, although he does use symbols—A's and B's and C's—to refer to differing determinants. Perhaps more surprising, he does not use diagrams to represent his ideas, and to deduce their consequences. The diagram mentioned above, that represents the consequences of segregation in meiosis, is a unique exception. The appearance of a complete theoretical biology, resting on mathematics, had to wait for Fisher, Haldane, and Wright.

Occasionally, Weismann's lack of mathematics lets him down, as in his treatment of normalizing selection and sex that was mentioned above, or in his attempted explanation of Gaussian distribution of phenotypes (2,206). Faced with the achievements of Darwin and Weismann, one cannot claim that mathematics, or even diagrams, are needed for successful theoretical work in biology. But they certainly make it easier.

7. WEISMANN TODAY

Many of the issues discussed by Weismann are still with us. I want to discuss only one, the central issue of the transmission of functional modifications. It is customary in elementary genetics to suggest that the issue is dead, but this is clearly not so. For Weismann, the difference between liver cells, kidney cells, fibroblasts, and so on, was caused by the presence within them of different determinants: only the germ track was totipotent. We now hold that, typically, somatic cells carry a complete diploid genome. Yet it remains true that differentiated cells breed true: fibroblasts give rise to fibroblasts, epithelial cells to epithelial cells, and so on. Usually the differences between such cells are in gene activation rather than in DNA sequence, but they are differences that can be transmitted to daughter cells at cell division, sometimes with a remarkable degree of stability. The heritable feature may be a pattern of methylation, which is replicated when DNA is replicated, but other features, including duplication of regions of the genome, are also possible. Thus there is a second inheritance system—an epigenetic inheritance system—in addition to the system based on DNA sequence that links sexual generations.

In an important article, Jablonka and Lamb (in press) review the cases in which changes initiated during ontogeny are known to be transmitted sexually, and discuss their evolutionary relevance. Two well-established examples are the 'imprinting' of mammalian chromosomes according to the sex of the parent, and the duplication of regions of the DNA in flax (*Linum*) caused by fertilizer treatments (Cullis 1983). I have not space here to discuss the evolutionary significance of such phenomena in any detail. I mention them only to show that the questions raised by Weismann are not dead, and to make one general theoretical point, which was already familiar to

Weismann. As explained above, when discussing the responses of cells to specific stimuli during development, Weismann pointed out that if the responses of cells to internal stimuli are such as to adapt the organism as a whole, then they must be explained by personal selection, and not by competition between cells. The same, I would argue, must be true in the context discussed by Jablonka and Lamb. If a heritable change is repeatably induced during ontogeny, is then transmissible sexually, and confers a selective advantage on the organism (in the two quoted examples, the last point has not been established, but is probably true), then the change must have been established by natural selection at the individual level. To clarify this point would require what Weismann was usually unwilling to provide, a formal model in diagrammatic form—but that is another story.

REFERENCES

Buss, L. W. (1987). *The evolution of individuality*. Princeton University Press. Princeton, New Jersey.

Cullis, C. A. (1983). Variable DNA sequences in flax. In *Genetic Rearrangement*, (ed. K. F. Chater *et al.*) pp. 253–64. Croom Helm, London.

Fisher, R. A. (1930). *The genetical theory of natural selection*. Oxford University Press.

Jablonka, E. and Lamb, M. J. (in press). The inheritance of acquired epigenetic variations. *J. Theor. Biol.* (in press).

Jacob, F. (1982). *The possible and the actual*. Pantheon, New York.

Kirkwood, T. B. L. and Cremer, T. (1982). Cytogerontology since 1881: a reappraisal of August Weismann and a review of modern progress. *Hum. Genet.* **60**, 101–21.

Kondrashov, A. S. (1982). Selection against harmful mutations in large sexual and asexual populations. *Genet. Res., Camb.* **40**, 325–32.

Mayr, E. (1988). On Weismann's growth as an evolutionist. In *Toward a new philosophy of biology*. Harvard University Press, Cambridge, Massachusetts.

Medawar, P. B. (1952). An unsolved problem in biology. Reprinted in *The uniqueness of the individual*. Methuen, London, 1957.

Muller, H. J. (1932). Some genetic aspects of sex. *Amer. Natur.* **66**, 118–38.

—— (1964). The relation of recombination to mutational advance. *Mutat. Res.* **1**, 2–9.

Weismann, A. (1891). *Essays upon heredity*, (2nd edn). Clarendon Press, Oxford.

—— (1904). *The evolution theory*, (trans. J. A. and M. R. Thomson). Edward Arnold, London.

Williams, G. C. (1957). Pleiotropy natural selection and the evolution of senescence. *Evolution* **11**, 398–411.

Life history variation in placental mammals: unifying the data with theory

PAUL H. HARVEY, ANDREW F. READ, and
DANIEL E. L. PROMISLOW

I. INTRODUCTION

There is considerable diversity in life histories among placental mammals. Litter size, for example, ranges from about 11 in the rodent *Tapera indica* through to the single young of many primates, ungulates, and cetaceans. To a reasonable approximation placental mammals can be arranged on a 'fast-slow continuum' (Eisenberg 1981; Stearns 1983), with small, short-lived, rapidly reproducing species like rodents at one end, and large species, such as elephants and whales, with the opposite suite of traits at the other end (Fig. 1). Closely related species tend to have similar life histories, and the largest differences are found among rather than within the major radiations (Eisenberg 1981; Table 1). The reasons for the evolution of this diversity are poorly understood. It is frequently claimed that species differences in some particular variable, such as brain size, body size, or metabolic rate constrain differences in reproductive parameters. We review the evidence for such claims, and argue that a more profitable research strategy is to examine differences in age-specific fertility and mortality schedules, which form the theoretical basis for understanding evolution in age-structured populations.

2. BODY SIZE

Bonner's (1965) famous bacterium to sequoia tree curve, which included mice and elephants, shows a linear relationship between the logarithms of generation time and body size for organisms ranging in size over almost eight orders of magnitude. Among placental mammals, such linear relationships between logarithmically transformed life history variables and body size (termed allometric relationships) have been reported for litter size, gestation length, duration of lactation, age at independence, and sexual maturity, and even for maximum lifespan recorded in zoos (Sacher 1959; Millar 1977, 1981; Blueweiss *et al.* 1978; Mace 1979; Western 1979; Western and Ssemakula

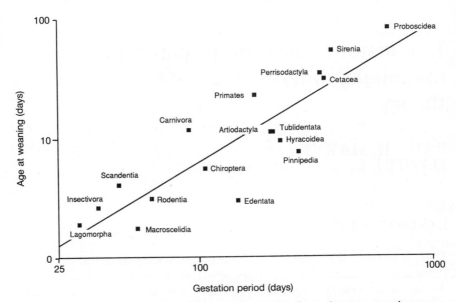

Fig. 1. Comparison between gestation period and age of weaning across order means for placental mammals. Order means calculated as described in Table 2. Orders containing large-bodied species have longer gestation periods and later ages at weaning. For details of data sources and taxonomy used see Table 1.

1982; Stearns 1983; Martin and MacLarnon 1985; Harvey and Clutton-Brock 1985; Gittleman 1986). Smaller-bodied species are short lived, but produce large litters of young which develop rapidly, whereas larger mammals reproduce and develop slowly and then live longer.

The potential differences in fecundity between different sized mammals can be remarkable. For example, consider the smallest and the largest species of primate. A female mouse lemur (*Microcebus murinus*) born at the same time as a female gorilla (*Gorilla gorilla*) could leave 10 million descendants before the gorilla became sexually mature, assuming equal reproductive success of the sexes. This difference results from the combined effects of differences in age at maturity, age at weaning, and litter size between mouse lemurs and gorillas (Harvey and Clutton-Brock 1985; Harvey *et al.* 1987).

It is not unusual for correlation coefficients describing the allometry of life history variables to be above 0.85 (see Harvey and Read 1988). Because of the historical interest in body size, and because correlations between life history variation and body size are so strong, the description of allometric relationships for life history variables among mammals has become something of an end in itself, with little attention being paid to their evolutionary causes. Instead, the assumption is often made—implicitly or explicity—that body size evolution constrains life history evolution, or that life history

Table 1

Taxonomic distribution of life history variance among placental mammals

Among- Within-	species genera	genera families	families orders	orders class
Gestation period	2(8)	6(15)	21(21)	71(56)
Age at weaning	8(18)	11(17)	19(20)	62(45)
Time from weaning to maturity	13(22)	12(20)	33(42)	42(16)
Age at maturity	11(24)	7(11)	27(35)	55(30)
Inter-litter interval	7(9)	14(14)	16(14)	64(63)
Maximum recorded lifespan	10(21)	10(24)	12(6)	68(49)
Neonate weight	3(34)	5(0)	27(33)	65(33)
Litter weight	3(46)	8(5)	29(33)	60(16)
Number of offspring per litter	8(18)	17(41)	8(6)	67(35)
Adult weight	3	7	21	69

Tabulated values are percentage of total variance accounted for at successive taxonomic levels. Numbers in parentheses are equivalent values for each variable after the effects of body size have been removed as described in text (p.16). Distribution of variance from nested ANOVA (Sokal and Rohlf 1981; Harvey and Mace 1982; Pagel and Harvey 1988b). Data are from 712 species from 18 orders extracted from the literature. Sources are given elsewhere (Read and Harvey 1989). Taxonomy follows Corbet and Hill (1980), Eisenberg (1981), and Nowak and Paradiso (1983a,b). Variables were defined as follows: gestation length as the average period from conception to birth, excluding periods of delayed implantation; age at weaning as period from birth to independence of the neonate from the maternal milk; inter-litter interval as the shortest period between births of successive litters.

variation is size-dependent (Blueweiss *et al.* 1978; Western 1979; Western and Ssemakula 1982; McMahon and Bonner 1983; Calder 1984; Schmidt-Nielsen 1984). Life histories are, under this view, claimed to be an inevitable consequence of some sort of anatomical and physiological scaling laws which are said to underlie the allometry.

In sharp contrast to our understanding of the mechanical principles behind many anatomical and physiological allometric relationships (Alexander 1968; Schmidt-Nielsen 1972, 1983; McMahon and Bonner 1983), the reasons for life history allometry remain virtually unknown. Yet without hypotheses about cause, allometric relationships are no more than restatements of the empirical facts they are purported to explain. Several authors have attempted to provide explanations for the scaling of life histories. For example, the concept of 'physiological time' (Brody 1945; Lindstedt and Calder 1981) has been extended to include the scale of life histories (Lindstedt and Calder 1981; Calder 1984), and even to the suggestion that there is a hidden 'periodengeber' (Lindstedt and Swain 1988). Such explanations are unsatisfactory for a variety of reasons.

First, there is considerable variation in most life history traits which is unaccounted for by body size. At about 150 kg, for example, lions *Panthera leo* produce on average 2.6 offspring after a gestation of 106 days, pigs *Sus scrofa* a litter of 6 after a gestation period of 120 days, and the dolphin *Tursiops truncatus* a singleton after a year-long gestation. Much of this variation is associated with taxonomic differences. For example, in most orders all species are altricial (eyes closed at birth) or all species are precocial (eyes open at birth). When altricial and precocial species of the same adult body size are compared, the former give birth to small neonates after a short gestation length while the latter give birth to large neonates after a long gestation (Martin 1984; Swihart 1984; May and Rubenstein 1985; Martin and MacLarnon 1985). Life history differences, therefore, are not just simple consequences of body size. Extending this argument, if life history allometry was derived from a physiologically imposed time scale, then other homeotherms, such as marsupials, birds, and monotremes of the same weight, might be expected to have the same life history tactics, which they do not (Western and Ssemakula 1982).

Secondly, although life history variables correlate well with body size, they often correlate even more closely with each other. For example, gestation length and age at weaning are highly correlated with each other independently of body size within primates (Harvey and Clutton-Brock 1985), Carnivora (Gittleman 1986), and Rodentia (Mace 1979). The same relationship holds across orders of placental mammals (Fig. 2). If the effects of body size are removed by calculating deviations from allometric relationships (to get relative values), then those orders with relatively long gestation length give birth to relatively small litters, relatively large neonates take relatively longer to wean and mature, and have relatively long potential lifespans (Table 2). Qualitatively, this is similar to the fast-slow continuum of life history strategies found when the influence of body size is not considered. However, the taxa fall into very different positions on the continuum when the effects of body size have been removed (compare Fig. 1 with Fig. 2). With body weight effects removed, rodents have developmental periods similar to those of cetaceans and artiodactyls (Fig. 2). Thus, size clearly has some effect. One explanation is that some of the factors responsible for the evolution of life history diversity in placental mammals are correlated with the size of the species, and that size may be merely a surrogate for them. We return to this suggestion below.

It is not our contention that size is unimportant in the evolution of the life histories among placental mammals. An elephant could not grow to reproductive size in the few weeks taken by a mouse. Nor could an elk produce the large neonate of an elephant. The question is not whether such constraints exist, but how important they are. For example, it is plausible that an upper limit on neonate size is sometimes set by the size of the birth canal or the ability of the placenta to sustain a large, growing fetus (Millar 1984). But, under the allometric perspective, it is by no means clear why elephants do not

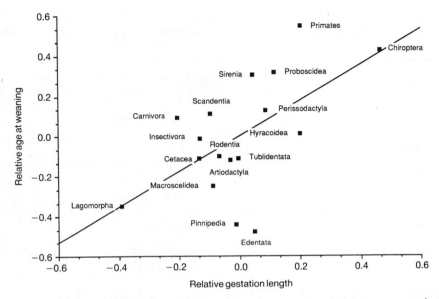

Fig. 2. Comparisons between relative gestation period and relative age at weaning across order means for placental mammals. Relative values refer to deviations from logarithmic regressions of gestation period and age at weaning on body weight, so that relative values are not correlated with body weight.

produce two horse-sized or three hippopotamus-sized neonates, which together would weigh the same as an elephant at birth (and fit through the birth canal more easily). The absence of certain trait combinations does not mean they could not arise. Furthermore, the large variation in life history tactics found within species, which is often unrelated to body size (Millar 1982; Boyce 1988), suggests that life histories are not simple consequences of body size. Of course why life histories covary with body size across species is a major challenge, but allometric arguments have little explanatory power until testable hypotheses are proposed about mechanism, and empirical support for them is provided.

The selective forces favouring differences in body size have received considerable attention in the past (Clutton-Brock and Harvey 1983) and many morphological, physiological, ecological, and behavioural factors covary with size (Clutton-Brock and Harvey 1979; Peters 1983; McMahon and Bonner 1983; Calder 1984). Size-related life history variation therefore has many potential explanations. Accordingly, our approach in the rest of this chapter is to focus on factors that are correlated with life history diversity independently of adult body size. Such factors may also covary with body size and, if they do, they may hold considerable promise as underlying variables influencing both size and life history.

Table 2

Correlation matrix of life history variables across order means after the effects of body size have been removed (as described in text)

	Gestation period	Age at weaning	Age at maturity	Inter-litter interval	Maximum recorded lifespan	Neonatal weight
Age at weaning	0.58* (17)					
Age at maturity	0.60* (15)	0.60* (15)				
Inter-litter interval	0.76† (14)	0.76† (13)	0.60* (13)			
Maximum lifespan	−0.31 (17)	0.43 (17)	0.30 (15)	0.65* (13)		
Neonate weight	0.63† (17)	0.21 (16)	0.33 (15)	0.24 (14)	0.24 (17)	
Offspring per litter	−0.63† (18)	−0.32 (17)	−0.49 (15)	−0.57* (14)	−0.69† (17)	−0.68† (17)

Number of orders given in parentheses. Means calculated for logarithmically transformed species values by calculating generic means, family means from the constituent generic means, and finally order means from family means. *$p<0.05$; †$p<0.01$. For details of the data and taxonomies on which the order means were calculated see Table 1.

3. METABOLIC RATE

It has been claimed that mammals may have reproductive rates which are as rapid as their metabolic rates allow (Millar 1977, 1984; Case 1978; Sacher 1978; McNab 1980, 1983, 1986; Western and Ssemakula 1982; Hennemann 1983; Hofmann 1983; Peters 1983; Calder 1984; Swihart 1984). The arguments concerning the reasons for species differences in metabolic rate are being debated elsewhere (Hennemann 1983; McNab 1983, 1987, 1988; Hayssen and Lacy 1985; Elgar and Harvey 1987; Harvey and Elgar 1987, Harvey et al., in press). If metabolic rates are important in life history evolution, then life history variation should be correlated with variation in metabolic rate when the effects of body size have been removed. Across and within the orders of placental mammals, this is not the case when basal metabolic rate is used as a measure of overall metabolic rate (Read and Harvey 1989; Harvey et al., in press). Therefore, the major differences in mammalian life histories cannot be attributed to basal metabolic rate independent of body size: basal metabolic rate has no more explanatory power than body size itself. Some other measure of metabolic rate, such as active daily metabolic rate, may be correlated with life history variation independent of body size, but no such relationship has yet been demonstrated even though it has been looked for (Harvey *et al.*, in press). However, even if there are no overall correlations between measures of metabolism and measures of fecundity, metabolism may be an important constraint in particular instances, such as among those species with very low metabolic rates relative to their body sizes.

4. BRAIN SIZE

It has been argued that the brain is a homeostatic organ and that animals with larger brains are consequently able to delay senescence (Sacher 1959). Furthermore, as neural tissue is slow to grow it may act as a limiting factor in development, so that larger-brained individuals take longer to develop (Sacher and Staffeldt 1974). These arguments were supported by early correlational evidence that both maximum recorded lifespan and average gestation length are more highly correlated with adult brain size than with adult body size (Sacher 1959; Sacher and Staffeldt 1974). More recent analyses do not always show stronger correlations of life history variation with brain than body size (Mace 1979; Bennet and Harvey 1985; Harvey and Clutton-Brock 1985; Gittleman 1986; Read and Harvey 1989). Brain and body size are so highly correlated in mammals (Harvey and Bennett 1983), that it often seems a matter of chance which variable is better correlated with life history variation independently of the effects of the other (Read and Harvey 1989). Furthermore, reduced intraspecific variation in brain sizes compared with body sizes may mean that brain size is less error-prone, in a statistical sense, than is body weight (Economos 1980): small samples may be composed of emaciated or obese individuals, while brain size varies little with body condition. Brain size may also be associated with some life history traits because of common correlations with other size-related variables (Lindstedt and Calder 1981; Bennett and Harvey 1985; Harvey and Clutton-Brock 1985; Harvey and Read 1988; Pagel and Harvey 1988*a*). As with basal metabolic rate, then, there is little empirical support for the view that adult brain size plays a more important causal role in life history evolution than does any other variable correlated with size.

5. ECOLOGY

A considerable body of theory has been developed by ecologists and evolutionary biologists, who view life histories as the result of natural selection acting on trade-offs imposed by the need to allocate finite resources between conflicting requirements, such as growth versus reproduction (Charlesworth 1980; Sibly and Calow 1986; Partridge and Harvey 1988). Trade-off curves are likely to differ among environments and, therefore, so are optimal life histories (Southwood 1988). The unification of this theory with the comparative data on placental mammals has proved difficult. The data have sometimes failed to support particular theoretical contentions (Millar 1984), but perhaps most discouraging has been the general absence of correlations between ecological variation and life history variation, once the effects of body size have been removed (Wootton 1987; Harvey and Read 1988). Perhaps this is not surprising for two reasons. First, differences in body size are correlated with differences in ecology, and differences in ecology are generally very crude estimates (for example, diet might be classified as leaves or fruit). Accordingly, any correlations between life histories and ecology that exist independently of size are unlikely to be evident once the primary correlations between ecology and size are removed. Second, most life history variation that is independent of size is found between major taxa (Table 1), yet species within those taxa tend to occupy a wide range of ecological niches.

One recent report does, however, demonstrate a relationship between fecundity and ecology independently of size. Primate species living in tropical rain forest have lower potential reproductive rates (Cole 1954) than species living in other habitats, such as savannah and secondary growth forest (Ross 1987). The same patterns can be detected across species within five of six taxonomic families represented in the sample (Harvey *et al.* 1989). Ross (1987) interprets this association as a consequence of r versus K selection. The r/K model is based on density dependence, and predicts that populations well below the carrying capacity for the environment (K) will be selected to have high r (intrinsic rate of increase) while those near the carrying capacity will be selected to have high K; despite the population selection connotation, the r/K model need not invoke group selection (Roughgarden 1971; Charlesworth 1980). Ross suggests that tropical rain forests are particularly stable environments inhabited by K-selected populations in which mortality is primarily density-dependent.

Following Pianka (1970), Ross goes on to argue that K selection favours longer periods of parental investment and delayed maturation which may be associated with the production of more competitive offspring. However, Pianka's (1970) assertions concerning r and K selection over extend the original model (Boyce 1984). It is important to distinguish between the empirically observed fast-slow continuum and the theoretically derived r/K continuum. For example, by definition, K selection results in increased carrying capacity, yet there is no good evidence that increased competitive ability among primates results in higher K (although, of course, it may). Indeed if selection did result in the carrying capacity being maximized, we might expect body sizes of primates living in tropical rain forests to be smaller than those living in other habitats. This is because smaller-bodied individuals have lower energetic requirements making them more successful in reproductive terms at indirect or scramble competition. A given amount of food can sustain more small than large-bodied mammals. In Ross' data there is no difference in body size between species from the different categories of habitat. Nevertheless, even if the labels r and K are inappropriate, Ross' explanation may still hold: there may be a trade-off between fecundity and competitive ability, and in tropical rain forests there may be a selective premium on competitive ability. More comprehensive discussions of the current status and limitations of r/K theory are given elsewhere (Boyce 1984, 1988; Partridge and Harvey 1988).

6. MORTALITY

Cole (1954) and Williams (1966) have pointed to the likely importance of age-specific mortality schedules on life history evolution. It is inevitable that some components of fecundity will be related to patterns of mortality because, in natural populations, mortality must be balanced by fecundity for a population to persist (Sutherland *et al.* 1986). However, there are many ways that components of life histories can be combined to produce the same reproductive rate. For example, decreased life expectancies could be balanced by changes in litter size, inter-birth interval, age at first reproduction, or some combination of all three. The challenge for comparative biologists is to explain which particular tactics are adopted and why.

A considerable body of theory shows how optimal fecundity schedules are likely to depend on age-specific mortality schedules (Cole 1954; Williams 1966; Murphy 1968; Gadgil and Bossert 1970; Charnov and Schaffer 1973; Schaffer 1974; Hirshfield and Tinkle 1975; Bell 1976; Wiley 1974; Horn 1978; Charlesworth 1980; Sibly and Calow 1986; Partridge and Harvey 1988). It is generally agreed that in stable age-structured populations, selection acts to maximize the Malthusian parameter or intrinsic rate of increase (r)—which here is the rate of increase of a population with a stable age distribution in a given environment—provided that certain assumptions hold (stable environment, frequency-independent weak selection, demographic ergodicity). The Malthusian parameter is determined by the schedules of age-specific survival and fertility characteristics of individuals in the population and is defined as:

$$1 = \int_0^w l_x m_x e^{-rx} dx$$

where x is age, l_x is the survival probability to age x, m_x is fertility at age x, and w is the last age of breeding. This theoretical framework may help elucidate variation in mammalian life history patterns. For example, the finding that primary species living in tropical rain forests have lower reproductive rates than those living in secondary forests and savannah (Ross 1987) might be the result of differences in age-specific survivorship associated with habitat (Harvey *et al.* 1989). For example, if adult mortality is high in the savannah (perhaps because of an increased risk of predation), then we might expect selection to favour higher reproductive effort even at the cost of some further adult mortality (Hirshfield and Tinkle 1975; Charlesworth 1980; Horn and Rubenstein 1984). Here, we are assuming stability of both habitats but different mortality schedules (l_x), whereas Ross assumed differences in environmental stability imposing a varying mortality schedule in one habitat.

We have used the association between primate life histories and habitat as an example of life history phenomena which might be explained by more than one theory. Clearly, efforts to relate different theories to the comparative fecundity data will be difficult until mortality schedules are documented. Recent studies using the relatively few mortality data that are available from placental mammals (Millar and Zammuto 1983; Harvey and Zammuto 1985; Sutherland *et al.* 1986; Harvey *et al.* 1989; Read and Harvey 1989, Promislow and Harvey, in press), like others from birds (Sæther 1988; Bennett and Harvey 1988), reveal striking and informative patterns: various measures of fecundity are correlated with mortality patterns independently of body size.

To take these analyses further, we have extracted data from the literature on mortality schedules for 48 placental mammal species living in approximately stable age-structured populations. We have compared these data with fecundity variables derived mainly from zoo populations, so that we are restricted to detecting evolutionary differences in fecundity among species rather than proximate responses to local population density. Ten different mammalian orders are represented in the sample. Only species in which age-specific mortality was recorded from birth to death of the oldest individual were included. In those species where the two sexes have appreciably different life histories or mortality patterns, we restricted our attention to the data from females. Larger-bodied species have lower mortality rates, both as juveniles and adults (Millar and Zammuto 1983; Harvey and Zammuto 1985), so to be certain that any relationships found are not just the result of common correlations, it is important to remove the effects of size from the analyses. Survivorship curves are well described by third order polynomial regressions when the logarithm of the number of survivors from a cohort is plotted against age (the average r^2 is 0.98).

$$\log(n_t) = \log(n_0) + b.t + c.t^2 + d.t^3$$

Where n_t refers to the number remaining in the cohort t years after birth. The best fit parameters b, c, and d are each significantly correlated with body weight ($p < 0.001$ in each case). Using a linear regression model, we were able to produce expected values for each parameter for a species of a given body size, thus enabling us to compare real mortality schedules with expected schedules based on body size for all species in the sample (for further details see Promislow and Harvey, in press). As a summary statistic, we measured the maximum deviation on a plot of number from cohort that were dead against age. This maximum deviation was positive when mortality was higher than expected. As the real data went into producing the expected figures (accounting for about 2 per cent of the variation as there were almost 50 species in the sample), deviations are slight underestimates of the true values.

Species with high mortality rates for their body size had the suite of life history characters associated with the fast end of the fast-slow continuum: short gestation lengths, early ages at weaning and maturity, short periods from weaning to maturity, and large litters (Table 3). Age at maturity, then, is not the only life history variable which changes with different mortality patterns. In particular, litter size, which has little effect on the intrinsic rate of increase relative to the contribution of age at maturity (Cole 1954; Lewontin 1965; Roughgarden 1971; Martin and Harvey 1987), is also strongly associated with mortality patterns. Furthermore, and not surprisingly, mortality correlates with body size independently of life history variation: small-bodied species have higher mortality rates (Kurtén 1953; Millar and Zammuto 1983; Harvey and Zammuto 1985). Body size may well have been acting in part as a surrogate measure for mortality in earlier analyses. If different ecologies

Table 3

Correlations across families of placental mammals between life histories and mortality with the effects of body size removed

	r	n	p
Gestation period	−0.44	21	0.048
Age at weaning	−0.50	18	0.040
Time from weaning to maturity	−0.52	18	0.026
Age at maturity	−0.61	20	0.004
Inter-litter interval	−0.47	20	0.037
Maximum recorded lifespan	−0.03	17	0.240
Neonate weight	−0.39	20	0.120
Litter weight	−0.26	20	0.260
Number of offspring per litter	+0.59	21	0.005

Data on age specific mortality were obtained from life tables of 48 species belonging to 21 families from 10 orders; sources are given elsewhere (Promislow and Harvey, in press). The summary mortality statistic used is described in text. Family means (calculated by averaging generic means) were used rather than order means because of the few orders for which data were available. r, partial product-moment correlation coefficient; n, number of families for which data on life history variable and mortality were available; P, probability level.

impose similar mortality schedules, these findings may help to explain the general failure to find relationships between ecology and life history variation when size effects are held constant.

The measure of mortality we have used is a summary statistic. However, theory predicts that the ratio of juvenile to adult morality will be a particularly relevant variable for understanding life history evolution (Cole 1954; Charnov and Schaffer 1973; Horn 1978; Sibly and Calow 1986). For example, if a perennial mammal has a probability of parental survival P and of offspring survival Y, an annual species (in which parental survival is reduced to zero) needs to add P/Y progeny to each brood in order to compensate. Unfortunately, very small biases in recording or recapture rates can result in artefactual correlations between juvenile and adult mortality rates (LeBreton 1977; Lakhani and Newton 1985; Anderson *et al.* 1985) and, in the absence of reliable estimates of bias, we are wary about using the available data to compare species differences in the ratio of juvenile to adult mortality rates.

Can we use the associations between our overall mortality statistic and fecundity variables to help explain the diversity in life histories of placental mammals? We take two examples. Here our intention is to put forward interpretations consistent both with the data currently available, and with some theoretical predictions. We hope to encourage the collection of the data necessary to test them, and the development of more rigorous predictions from theory.

Mammals with long life expectations delay reproduction to beyond an age when other species of the same body size have already started to reproduce (Harvey and Zammuto 1985; Sutherland *et al.* 1986). If there are mortality costs associated with reproduction, and if reproductive efficiency increases with age, then animals with lower non-reproductive mortality rates should be selected to delay reproduction (Ashmole 1963; Charlesworth 1980; Harvey 1989). However, species with high non-reproductive mortality rates should be selected to reproduce early because they are less likely to survive to reproduce when reproductive efficiency is high. For example, if females die after reproduction, and can produce one offspring in their first year of life or five in their second year they will be selected to wait until their second year to reproduce if, and only if, non-reproductive mortality is low enough. If the chances of survival to their second year are, say 1 per cent, they would do better to reproduce in their first year.

The assumption, then, is that reproductive efficiency in mammals increases with age. The evidence for this is reasonably good: reproduction among older, but not senescent animals (in which reproductive effort might be increased for other reasons), is associated with larger young (Fleming and Rauscher 1978; Myers and Master 1983; Calambodkis and Gentry 1985; Stewart 1986), larger litters (Fleming and Rauscher 1978; Allen 1983, 1984; Dunn and Chapman 1983; Myers and Master 1983; Sæther 1983; Verme 1983; Payne 1984; Von María *et al.* 1985), higher survival rates of the young

(Drickamer 1974; Fleming and Rauscher 1978; Allen 1984; Calambodkis and Gentry 1985), and shorter inter-birth intervals (Clutton-Brock *et al.* 1982). Some of these patterns may result in part from experience of older parents, and experiments are necessary to evaluate the effects of experience and age. The predicted trade-off is between delaying reproduction to a time when reproductive efficiency is high and dying while waiting. A similar argument for the evolution of delayed reproduction has been made for sea birds (Ashmole 1963). We wonder how much of the variation in age at maturity which Wootton (1987), for example, claims is 'explained' by body size is, in turn, explained by size-dependent variation in non-reproductive mortality rates.

Mammals of similar body sizes may give birth to poorly developed altricial young after a short gestation length, or advanced precocial young after a long gestation (Sacher and Staffeldt 1974; Case 1978; Eisenberg 1981; Pagel and Harvey 1988a). It has been suggested that one of the advantages of producing precocial young is that 'if the young survives, the relative proportion of parental care may be reduced' (Eisenberg 1981). We should then predict a negative relationship between the periods of prenatal invest-ment (gestation length) and postnatal investment (age at weaning or independence) after the effects of body mass have been removed by partial correlation. In fact, as we have already mentioned, the relationship is positive within and among orders of mammals: those mammals with long gestation lengths produce developmentally advanced young which wean late. Why should the periods of prenatal and postnatal investment be positively correlated independently of size? Perhaps, again, we should look to morta-lity? If a mother dies before her offspring are weaned in the wild, the young will also perish, thus favouring shorter periods of parental investment. If long periods of investment lead to higher quality offspring, where quality is measured in terms of survival or fecundity, in species with low rates of mortality, individuals which invest for longer in their young both prenatally and postnatally will be favoured, a prediction confirmed by the available data (Table 3). Similar positive relationships between the duration of different periods of maternal investment and mortality, after size effects have been removed, have been found in birds (Lack 1968; Sæther 1988; Bennett and Harvey 1988), and may be similarly explained.

The explanations given above for species differences in the age at first reproduction and patterns of parental investment by mammals result in an exaggeration of differences in mortality rates between species. If species with high rates of non-reproductive mortality start reproducing when they are younger and wean their young earlier, we would expect mortality rates of both young and adults to be further increased. Other examples of such self-reinforcing or positive feedback systems in life history evolution are de-scribed elsewhere (Horn 1978; Horn and Rubenstein 1984). It might seem tempting to suggest that the altricial/precocial dichotomy mentioned above arises as a result of species being selected to follow one or other life history strategy, but this view depends on an oversimplification: the altricial/ precocial dichotomy can be shown to exist independently of the fast-slow continuum discussed in this chapter (e.g. Stearns 1983).

7. CONCLUSION

We are only too well aware of the preliminary nature of the conclusions reached in this article. In defence, we echo Tukey's (1962) words 'Far better an approximate answer to the right question, which is often vague, than an exact answer to the wrong question, which can always be made precise'. We believe that Williams (1966) was asking the right question over 20 years ago when he claimed that 'relating phylogenetic variation in reproductive functions with phylogenetic variation in demographic factors of age-specific fecundity and death rate will reveal much of explanatory value'. Perhaps, in the intervening period many comparative biologists have lost sight of Williams' vision.

8. SUMMARY

Traditional interpretations of life history diversity in placental mammals view reproductive rates primarily as a consequence of species differences in body size, brain size or metabolic rate. These ideas have little theoretical basis and are unsupported by empirical evidence. A more profitable approach combines optimality theory with comparative data on fertility and mortality schedules which vary with habitat. Viewed in this way, differences in life histories among placental mammals may be explained in terms of evolutionary models.

ACKNOWLEDGEMENTS

We thank A. Burt, A. E. Keymer, M. D. Pagel, S. Nee, and S. J. Stearns for access to data, helpful discussions, or comments on the draft manuscript. A.F.R. was financed by a Commonwealth Scholarship from the Association of Commonwealth Universities and by Christ Church, Oxford. D.E.L.P was financed by the Rhodes trust. This chapter reports part of a research programme funded by Contract Number SC1*.0184.C from the Commission of the European Communities.

REFERENCES

Alexander, R. M. (1968). *Animal mechanics*. University of Washington Press, Seattle.

Allen, S. H. (1983). Comparison of red fox litter sizes determined from counts of embryos and placental scars. *J. Wildl. Manag.* **49**, 860–2.

——(1984). Some aspects of reproductive performance in female red fox in North Dakota. *J. Mammal.* **65**, 246–55.

Anderson, D. R., Burnham, K. P., and White, G. C. (1985). Problems in estimating age-specific survival rates from recovery data of birds ringed as young. *J. Anim. Ecol.* **54**, 89–98.

Ashmole, N. P. (1963). The regulation numbers of oceanic birds. *Ibis* **103b**, 458–73.

Bell, G. (1976). On breeeding more than once. *Amer. Natur.* **110**, 57–77.

Bennett, P. M. and Harvey, P. H. (1985). Relative brain size and ecology in birds. *J. Zool.* **207**, 491–509.

—— and ——(1988). How fecundity balances mortality in birds. *Nature* **333**, 216.

Blueweiss, L., Fox, H., Kudza, V., Nakashima, D., Peters, R. H., and Sams, S. (1978). Relationships between body size and some life history parameters. *Oecologia* **37**, 257–72.

Bonner, J. T. (1965). *Size and cycle: an essay in the structure of biology*. Princeton University Press, Princeton, New Jersey.

Boyce, M. S. (1984). Restitution of r- and K-selection as a model of density dependent natural selection. *Ann. Rev. Ecol. Syst.* **15**, 427–48.

——(ed.) (1988). *Evolution of life histories: pattern and theory from mammals*. Yale University Press, New Haven, Connecticut.

Brody, S. (1945). *Bioenergetics and growth*. Hafner, New York.

Calambokidis, J. and Gentry, R. L. (1985). Mortality in northern fur seal populations in relation to growth and birth weights. *J. Wildl. Dis.* **21**, 327–30.

Calder, W. A. (1984). *Size, function and life history*. Harvard University Press, Cambridge, Massachusetts.

Case, T. J. (1978). On the evolution and adaptive significance of post-natal growth in terrestrial vertebrates. *Q. Rev. Biol.* **53**, 243–82.

Charlesworth, B. (1980). *Evolution in age-structured populations*. Cambridge University Press.

Charnov, E. L. R. and Schaffer, W. M. (1973). Life history consequences of natural selection: Cole's result revisited. *Amer. Natur.* **107**, 791–3.

Clutton-Brock, T. H. and Harvey, P. H. (1979). Comparison and adaptation. *Proc. Roy. Soc. Lond.* B **205**, 547–65.

—— and ——(1983). The functional significance of body size variation among mammals. In *Advances in the study of animal behavior* (ed. J. F. Eisenberg and D. G. Kleiman), pp. 632–63. American Society of Mammalogists, New York.

——, Guinness, F. E., and Albon, S. D. (1982). *Red deer: behavior and ecology of two sexes*. Chicago University Press, Chicago.

Cole, L. C. (1954). The population consequences of life history phenomena. *Q. Rev. Biol.* **29**, 103–37.

Corbet, G. B. and Hill, J. E. (1980). *A world list of mammalian species*. British Museum (Natural History), London.

Drickamer, L. C. (1974). A ten-year study of reproductive data for free ranging *Macaca mulatta*. *Folia Primat.* **21**, 61–80.

28   Paul H. Harvey, Andrew F. Read, and Daniel E. L. Promislow

Dunn, J. P. and Chapman, J. A. (1983). Reproduction, physiological responses, age structure and food habits of racoon in Maryland, USA. *Z. Saug.* **48**, 161–75.

Economos, A. C. (1980). Brain-lifespan conjecture: a re-evaluation of the evidence. *Geront.* **26**, 82–9.

Eisenberg, J. F. (1981). *The mammalian radiations.* Athlone Press, London.

Elgar, M. A. and Harvey, P. H. (1987). Basal metabolic rates in mammals: allometry, phylogeny and ecology. *Funct. Ecol.* **1**, 25–36.

Fleming, T. H. and Rauscher, R. H. (1978). On the evolution of litter size in *Peromyscus leucopus. Evolution* **32**, 45–55.

Gadgil, M. and Bossert, W. H. (1970). Life historical consequences of natural selection. *Amer. Natur.* **104**, 1–244.

Gittleman, J. L. (1986). Carnivore life history patterns: allometric, phylogenetic and ecological associations. *Amer. Natur.* **127**, 744–71.

Harvey, P. H. (in press). Life history variation: size and mortality patterns. In *Primate life histories and evolution,* (ed. C. J. De Rousseau). Alan R. Liss, New York.

—— and Bennett, P. M. (1983). Brain size, energetics, ecology and life history patterns. *Nature* **306**, 314–15.

—— and Clutton-Brock, T. H. (1985). Life history variation in primates. *Evolution* **39**, 559–81.

—— and Elgar, M. A. (1987). In defence of the comparative method. *Funct. Ecol.* **1**, 160–1.

—— and Mace, G. M. (1982). Comparisons between taxa and adaptive trends: problems of methodology. In *Current problems in sociobiology* (ed. King's College Sociobiology Group), pp. 343–61. Cambridge University Press.

—— and Read, A. F. (1988). How and why do mammalian life histories vary? In *Evolution of life histories: pattern and process from mammals* (ed. M. S. Boyce), pp. 213–32. Yale University Press, New Haven, Connecticut.

—— and Zammuto, R. M. (1985). Patterns of mortality and age at first reproduction in natural populations of mammals. *Nature* **315**, 319–20.

——, Martin, R. D., and Clutton-Brock, T. H. (1987). Primate life histories in comparative perspective. In *Primate societies* (ed. B. B. Smuts, D. L. Cheney, R. M. Seyfarth, R. W. Wrangham, and T. T. Struhsaker), pp. 181–96. Chicago University Press, Chicago.

——, Promislow, D. E. L., and Read, A. F. (1989). Causes and correlates of life history differences among mammals. In *Comparative socioecology* (ed. V. Standen and R. Foley), pp. 305–18. Blackwell, Oxford.

——, Rees, J. A., and Pagel, M. D. (in press). Mammal metabolism and life histories. *Amer. Natur.*

Hayssen, V. and Lacy, R. C. (1985). Basal metabolic rates in mammals: taxonomic differences in the allometry of BMR and body mass. *Comp. Biochem. Physiol.* **81**, 741–54.

Hennemann, W. W. (1983). Relationship among body mass, metabolic rate and the intrinsic rate of increase in mammals. *Oecologia* **56**, 104–8.

Hirshfield, M. F. and Tinkle, D. W. (1975). Natural selection and the evolution of reproductive effort. *Proc. Natl. Acad. Sci. USA* **72**, 2227–31.

Hofmann, M. A. (1983). Energy metabolism, brain size and longevity in mammals. *Q. Rev. Biol.* **58**, 495–512.

Horn, H. S. (1978). Optimal tactics of reproduction and life history. In *Behavioural ecology: an evolutionary approach* (ed. J. R. Krebs and N. B. Davies), pp. 272–94. Blackwell, Oxford.

——and Rubenstein, D. I. (1984). Behavioural adaptations and life history. In *Behavioural ecology: an evolutionary approach*, (2nd edn) (ed. J. R. Krebs and N. B. Davies), pp. 279–98. Blackwell, Oxford.

Kurtén, B. (1953). On the variation and population dynamics of fossil and recent mammal populations. *Acta Zool. Fenn.* **76**, 1–122.

Lack, D. (1968). *Ecological adaptations for breeding in birds.* Methuen, London.

Lakhani, K. H. and Newton, I. (1985). Estimating age-specific bird survival rates from ringing recoveries—can it be done? *J. Anim. Ecol.* **52**, 83–91.

LeBreton, J. D. (1977). Multinomial method to estimate survival rate from bird band return: some complements to the age dependent method. *Biometrie-Praximetrie* **17**, 145–61.

Lewontin, R. C. (1965). Selection for colonizing ability. In *The genetics of colonizing species* (ed. H. G. Baker and G. L. Stebbins), pp. 79–94. Academic Press, New York.

Lindstedt, S. L. and Calder, W. A. (1981). Body size, physiological time and longevity of homeothermic animals. *Q. Rev. Biol.* **56**, 1–16.

——and Swain, S. D. (1988). Body size as a constraint of design and function. In *Evolution of life histories of mammals: theory and pattern* (ed. M. S. Boyce), pp. 93–106. Yale University Press, New Haven, Connecticut.

Mace, G. M. (1979). The evolutionary ecology of small mammals. D. Phil. thesis. University of Sussex, Brighton.

Martin, R. D. (1984). Scaling effects and adaptive strategies in mammalian lactation. *Symp. Zool. Soc. Lond.* **51**, 81–117.

——and Harvey, P. H. (1987). Human bodies of evidence. *Nature* **330**, 697–8.

——and MacLarnon, A. M. (1985). Gestation length, neonatal size and maternal investment in placental mammals. *Nature* **313**, 220–3.

May, R. M. and Rubenstein, D. I. (1984). Reproductive Strategies. In *Reproduction in mammals 4: reproductive fitness* (ed. C. R. Austin and R. V. Short), pp. 1–23. Cambridge University Press.

McMahon, T. A. and Bonner, J. T. (1983). *On size and life.* Scientific American, New York.

McNab, B. K. (1980). Food habits, energetics and population biology of mammals. *Amer. Natur.* **116**, 106–24.

——(1983). Energetics, body size, and the limits to endothermy. *J. Zool.* **199**, 1–29.

——(1986). The influence of food habits on the energetics of eutherian mammals. *Ecol. Monogr.* **56**, 1–19.

——(1987). Basal rate and phylogeny. *Funct. Ecol.* **1**, 159–60.

——(1988). Complications inherent in scaling the basal rate of metabolism of mammals. *Q. Rev. Biol.* **63**, 25–53.

Millar, J. S. (1977). Adaptive features of mammalian reproduction. *Evolution* **31**, 370–86.

——(1981). Pre-partum reproductive characteristics of eutherian mammals. *Evolution* **35**, 1149–63.

——(1982). Life cycle characteristics of northern *Peromyscus maniculatus borealis.* *Canad. J. Zool.* **60**, 510–15.

—— (1984). The role of design constraints in the evolution of mammalian reproductive rates. *Acta Zool. Fenn.* **171**, 133–6.
—— and Zammuto, R. M. (1983). Life histories of mammals: an analysis of life
tables. *Ecology* **64**, 631–5.
Murphy, G. I. (1968). Pattern in life history and the environment. *Amer. Natur.* **102**,
391–403.
Myers, P. and Master, L. L. (1983). Reproduction by *Peromyscus maniculatus*: size
and compromise. *J. Mammal.* **64**, 1–18.
Nowak, R. L. and Paradiso, J. L. (1983*a*). *Walker's mammals of the world: Volume 1.*
Johns Hopkins University Press, Baltimore.
—— and —— (1983*b*). *Walker's mammals of the world: Volume 2.* Johns Hopkins
University Press, Baltimore.
Pagel, M. D. and Harvey, P. H. (1988*a*). How mammals produce large-brained
offspring. *Evolution* **42**, 948–57.
—— and —— (1988*b*). Recent developments in the analysis of comparative data. *Q.
Rev. Biol.* **63**, 413–40.
Partridge, L. and Harvey, P. H. (1988). The ecological context of life history
evolution. *Science* **241**, 1449–55.
Payne, N. F. (1984). Reproductive rates of beaver in Newfoundland. *J. Wildl. Manag.*
48, 912–17.
Peters, R. H. (1983). *The ecological implications of body size.* Cambridge University
Press.
Pianka, E. R. (1970). On *r* and *K* selection. *Amer. Natur.* **104**, 592–6.
Promislow, D. E. L. and Harvey, P. H. (in press). Living fast and dying young: a
comparative analysis of life history variation among mammals. *J. Zool.*
Read, A. F. and Harvey, P. H. (1989). Life history differences among the eutherian
radiations. *J. Zool.*
Ross, C. R. (1987). The intrinsic rate of natural increase and reproductive effort in
primates. *J. Zool.* **214**, 199–220.
Roughgarden, J. (1971). Density-dependent natural selection. *Ecology* **52**, 453–68.
Sacher, G. A. (1959). Relationship of lifespan to brain weight and body weight in
mammals. In *CIBA Foundation Symposium on the Lifespan of Animals* (ed. G. E.
W. Wolstenholme and M. O'Connor), pp. 115–33. Little, Brown and Co., Boston,
Massachusetts.
—— (1978). Longevity and ageing in vertebrate evolution. *Bioscience* **28**, 297–301.
—— and Staffeldt, E. F. (1974). Relationship of gestation time to brain weight for
placental mammals. *Amer. Natur.* **108**, 593–616.
Sæther, B. E. (1983). Life history of the moose. *J. Mammal.* **64**, 226–32.
—— (1988). Evolutionary adjustment of reproductive traits to survival rates in
European birds. *Nature* **331**, 616–17.
Schaffer, W. M. (1974). Selection for optimal life histories: the effects of age structure.
Ecology **55**, 291–303.
Schmidt-Nielsen, K. (1972). *How animals work.* Cambridge University Press.
—— (1983). *Animal physiology. Adaptation and environment.* Cambridge University
Press.
—— (1984). *Scaling. Why is animal size so important?* Cambridge University Press.
Sibly, R. M. and Calow, P. (1986). *Physiological ecology of animals: an evolutionary
approach.* Blackwell, Oxford.
Sokal, R. R. and Rohlf, F. J. (1981). *Biometry.* Freeman, New York.

Southwood, T. R. E. (1988). Tactics, strategies and templets. *Oikos* **52**, 3–18.

Stearns, S. C. (1983). The influence of size and phylogeny on life history patterns. *Oikos* **41**, 173–87.

Stewart, R. E. A. (1986). Energetics of age specific reproductive effort in female harp seals *Phoca groenlandicus*. *J. Zool.* **208**, 503–17.

Sutherland, W. J., Grafen, A., and Harvey, P. H. (1986). Life history correlations and demography. *Nature* **320**, 88.

Swihart, R. K. (1984). Body size, breeding season length and life history tactics of lagomorphs. *Oikos* **43**, 282–90.

Tukey, J. W. (1962). The future of data analysis. *Ann. Math. Stat.* **33**, 1–67.

Verme, C. (1983). Sex ratio variation in *Odocoileus*: a critical review. *J. Wildl. Manag.* **47**, 573–82.

Von María, J. L., Gosálbez, J., and Sans-Coma, V. (1985). Über die Fortpflanzung der Hausspitzmaus (*Crocidura russula* Hermann, 1780) im Ebro-Delta (Katalonien, Spanien). *Z. Saug.* **50**, 1–6.

Western, D. (1979). Size, life history and ecology in mammals. *Afr. J. Ecol.* **17**, 185–204.

—— and Ssemakula, J. (1982). Life history patterns in birds and mammals and their evolutionary interpretation. *Oecologia* **54**, 281–90.

Wiley, R. H. (1974). Effects of delayed reproduction on survival, fecundity and rate of population increase. *Amer. Natur.* **108**, 705–9.

Williams, G. C. (1966). *Adaptation and natural selection.* Princeton University Press, Princeton, New Jersey.

Wootton, J. T. (1987). The effects of body mass, phylogeny, habitat, and trophic level on mammalian age at first reproduction. *Evolution* **41**, 732–49.

Comparative methods using phylogenetically independent contrasts

AUSTIN BURT

I. INTRODUCTION

As the data base of biology grows at its ever increasing pace, comparisons among species are becoming an increasingly popular means of testing evolutionary and ecological theory. Among the more common analyses are tests of association between two variables across some defined set of species; the variables of interest may be properties of individuals (e.g. testis size, Harcourt *et al.* 1981), of populations (e.g. genetic polymorphism, Nevo *et al.* 1984), or of the whole species (e.g. number of pests, Strong *et al.* 1984). Concurrent with this trend has been a growing appreciation of the statistical properties of such comparisons, in particular the 'problem of phylogeny', and a widening array of statistical methods from which a comparative biologist may choose (e.g. Harvey and Mace 1982; Ridley 1983; Clutton-Brock and Harvey 1984; Felsenstein 1985; Bell 1989; Grafen 1989; see review by Pagel and Harvey 1988*a*). In this Chapter, I try to put some of the issues into perspective and to expand upon one particularly promising approach to working with comparative data. I first discuss in detail the consequences of phylogenetic similarity—the common observation that closely related species are more similar than distantly related species—for the inferences one may draw from species-level correlations. These considerations lead to the notion of dividing a phylogeny into a series of independent replicate comparisons; these 'contrasts' may then be used in tests of association which are more discriminating—and thus often more interesting—than species-level analyses. They may also be used in conjunction with information on times of divergence in the evolutionary tree to address a number of explicitly historical topics, such as the relationship between speciation and phenotypic change, the suitability of random walk models of evolution, and the notion of phylogenetic inertia. I shall illustrate all analyses with the data of Sessions and Larson (1987) on DNA content and differentiation rate in plethodontid salamanders.

2. PHYLOGENETIC SIMILARITY

If one has an hypothesis that X plays some role in determining Y, then one obvious test is to look for a correlation between X and Y, using each species for which data are available as an independent data point. However, this practice has recently been the target of some extensive criticism (e.g. Harvey and Mace 1982; Ridley 1983; Felsenstein 1985; Grafen (1989), although accounts differ in identifying the exact source of the problem. Here I make use of results from the computer simulations of Raup and Gould (1974). These authors generated a phylogenetic tree from a single ancestral species by an iterative process of random speciation and extinction, until there were 200 descendants. Phenotypic evolution of 10 hypothetical characters was then modelled by 10 independent random walks: each character started from an ancestral value of zero and evolved by randomly increasing or decreasing one unit (or staying the same) at each branch point. Raup and Gould (1974) then calculated the correlation coefficient for all 45 pairwise combinations of the 10 variables. As can be seen in Fig. 1, the frequency distribution of their correlation coefficients is much wider than that expected from 200 pairs of normally distributed random variables—indeed, fully 45 per cent of their correlation coefficients are significant by the conventional test, despite the fact that the characters evolved by independent random walks. This difference between the observed distribution of correlation coefficients and the conventional theoretical distribution is due to phylogenetic similarity: because descendant phenotypes are modifications of ancestral phenotypes, closely related species are likely to be more similar than distantly related species. Significant correlations arise because species with similar X-values are likely to be closely related and thus also likely to have similar Y-values, independent of any causal mechanism relating changes in X to changes in Y. Note that it does not really matter *why* closely related species are more similar than distantly related species—be it because of adaptation to common environments, phylogenetic inertia, developmental constraints, or whatever—but as long as they are, then X and Y will tend to be correlated. These simulations demonstrate that significant interspecific associations between two variables may be due not only to causal mechanisms, simple or complex, but also to the rather common (and perhaps boring) tendency for closely related species to be more similar than distantly related species. A statistical test which could distinguish between these possibilities would be very useful.

Before describing such a test, a few further comments on species-level analyses seem appropriate. Consider Raup and Gould's 200 simulated descendant species as an order of mammals with parametric correlation coefficient between X and Y equal to ρ. If one were to sample randomly from this population and calculate the correlation coefficient and associated probability level for this sample, these statistics would have many of the

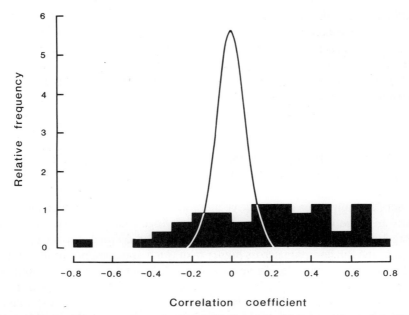

Correlation coefficient

Fig. 1. The problem of phylogenetic similarity. The histogram is a frequency distribution of correlation coefficients from the computer simulations of Raup and Gould (1974, table 2). Phenotypic evolution was modelled as independent random walks along a simulated phylogenetic tree with 200 descendent species. Forty-five pairwise correlation coefficients are shown, as well as the expected distribution of correlation coefficients for $n = 200$ pairs of normally distributed random variables (normal distribution with mean $= 0$ and variance $= 1/198$; Sokal and Rohlf 1981, p. 583). The correlation coefficients are from only one tree and are all pairwise combinations of only 10 variables; nevertheless, the sizeable difference between null models is clear. In particular, note that the distribution of simulated correlation coefficients is much wider than the theoretical curve, resulting in many statistically significant values under the conventional tests of association, despite the fact that the characters evolved by independent random walks.

conventional interpretations: the sample correlation coefficient, r, would still be a good estimate of ρ and the p-value would still give the probability that the parametric ρ has sign opposite to that of the sample r. For some purposes these statistics may be all that is required, with rather little depending on whether an association is causal or due to phylogenetic similarity. On the other hand, the simulations of Raup and Gould indicate that these statistics may be a very poor indicator of the correlation in another order of mammals, even if the same biological processes were occurring—if associations are completely due to phylogenetic similarity then a significant positive association in one order is just as likely to be negative in another. Certainly, inferring causality from a correlation (i.e. extrapolating to the outcome of manipulative experiments) will be even more tenuous than usual if there is a good chance that the correlation is due to phylogenetic similarity. To conclude, species-level analyses are not 'wrong', and may in fact reveal much of interest; however, they do permit fewer inferences than other more powerful techniques.

3. PHYLOGENETICALLY INDEPENDENT CONTRASTS AND TESTS OF ASSOCIATION

There are numerous comparative methods in the literature which attempt to deal in some way with the problem of phylogenetic similarity (reviewed in Pagel and Harvey 1988a); briefly, these may be divided into three groups. First, one can compare across higher taxa, such as genera or families, instead of species. For example, Harvey and Clutton-Brock (1985) use subfamilies in their analysis of primate life history data. Secondly, one can control for taxonomic affiliation directly by including it as a categorical variable in a multiple regression model. The methods used by Stearns (1983) and Wootton (1987) in their analyses of mammalian life history data are elaborations on this theme. Computer simulations by Grafen (1989) suggest that while using either of these methods does reduce the likelihood that a significant association is due to phylogenetic similarity, neither one completely excludes this alternative explanation. Finally, one can divide the data set into a series of phylogenetically independent replicate comparisons, each of which contributes one degree of freedom to the subsequent test of association. It is this latter approach which I will pursue here.

Consider again the simulations of Raup and Gould (1974). In any one particular simulation the correlation between two variables may be highly significant under the conventional test. However, in repeated simulations the expected correlation is zero—sometimes 'significantly' positive, sometimes negative, but with a long-term expected mean of zero. In theory, then, one could test for an association between two variables in an order of mammals by evolving the order repeatedly and seeing if there is a tendency for positive or negative correlation coefficients to predominate. In practice, one may proceed in a completely analogous manner: divide the order into independent replicate groups containing at least two species and then see if there is a tendency for positive or negative trends within groups to predominate. This is the method of phylogenetically independent contrasts.

Replicating comparisons within a taxon is not a particularly new idea, although it has a far from consistent history in the comparative literature. For example, to test for an association between rates of recombination and breeding system, Brown (1961) compares chiasma frequencies within six pairs of closely related species or subspecies in the genus *Gilia* (Polemonia-ceae), where each pair consists of a selfer and an outcrosser. More recently, Felsenstein (1985, p. 13) cites essentially the same suggestion: one should compare pairs of nearest relatives, 'two seals, two whales, two bats, two deer, etc.' Read (1987) takes a slightly different approach in his test of association between parasite load and plumage coloration in passerines, calculating Spearman rank correlations separately for each genus in his data set and then testing for mean different from zero. Here I extend these methods to the problem of extracting independent contrasts from a known phylogeny; Felsenstein (1988, p. 457) independently describes much the same method. First, however, I introduce the example data set.

Simple biochemical considerations suggest that if most of the interspecific variation in genome size is in the non-coding fraction then the more DNA there is, the longer will be the cell cycle and the slower the rate of development. These ideas suggest that there will be a negative correlation between DNA content and developmental rate across species; to test this prediction, Sessions and Larson (1987) collected data on the *C*-value (haploid DNA content) and the differentiation rate of regenerating limbs for 27 species of plethodontid salamanders. These data are reproduced in Fig. 2. (See Sessions and Larson (1987) for more details on theories and variables.) There is indeed a significant negative correlation between *C*-value and differentiation rate (Fig. 3a) and the parametric correlation coefficient for the Plethodontidae seems to be quite low (assuming this is a random sample, 95 per cent confidence limits are -0.69 and -0.93). Thus, the causal hypothesis gains some measure of support. However, this test does not reject the null hypothesis that *C*-value and differentiation rate have evolved independently of one another: as demonstrated above, such associations are not improbable if there is considerable phylogenetic similarity. This alternative explanation is supported by an analysis of variance, which indicates very highly significant differences between the two subfamilies in the data set in both *C*-value and differentiation rate ($F_{1,25} = 21.9$ and 45.3 respectively, both $p < 0.001$). Estimates of the variance components indicate that over 75 per cent of the total variance in both *C*-value and differentiation ,rate lies between subfamilies; these considerable differences are also evident from the scatterplot (Fig. 3a). Thus, closely related species are much more likely to be similar than distantly related species—is this fact alone responsible for the observed correlation?

To test further the causal hypothesis I divide the data set into a series of replicate contrasts. A contrast is simply a group of two or more species; it can be represented as a path joining the member species through phylogeny. I suggest that in tests of association one only use contrasts whose paths do not meet at any point; contrasts which satisfy this criterion are 'phylogenetically independent'. Note that variances and covariances of phenotypic characters within contrasts, which are assumed to depend only on evolutionary events occurring since the last common ancestor, will thereby be independent of

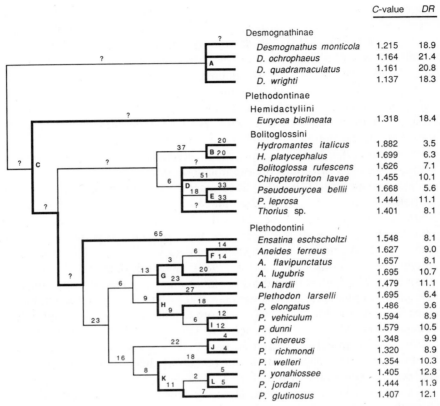

	C-value	DR
Desmognathinae		
Desmognathus monticola	1.215	18.9
D. ochrophaeus	1.164	21.4
D. quadramaculatus	1.161	20.8
D. wrighti	1.137	18.3
Plethodontinae		
Hemidactyliini		
Eurycea bislineata	1.318	18.4
Bolitoglossini		
Hydromantes italicus	1.882	3.5
H. platycephalus	1.699	6.3
Bolitoglossa rufescens	1.626	7.1
Chiropterotriton lavae	1.455	10.1
Pseudoeurycea bellii	1.668	5.6
P. leprosa	1.444	11.1
Thorius sp.	1.401	8.1
Plethodontini		
Ensatina eschscholtzi	1.548	8.1
Aneides ferreus	1.627	9.0
A. flavipunctatus	1.657	8.1
A. lugubris	1.695	10.7
A. hardii	1.479	11.1
Plethodon larselli	1.695	6.4
P. elongatus	1.486	9.6
P. vehiculum	1.594	8.9
P. dunni	1.579	10.5
P. cinereus	1.348	9.9
P. richmondi	1.320	8.9
P. welleri	1.354	10.3
P. yonahiossee	1.405	12.8
P. jordani	1.444	11.9
P. glutinosus	1.407	12.1

Fig. 2. The data of Sessions and Larson (1987) on C-value (haploid DNA content in pg, log-transformed) and differentiation rate of regenerating hind limbs (units are (developmental stages per day) × 100), in plethodontid salamanders. To the left are phylogenetic relationships, inferred from morphological and biochemical data, which are used in identifying independent contrasts. Contrasts are groups of at least two species and may be represented as a path joining the member species through the phylogeny; two or more paths are phylogenetically independent if their paths do not meet or fall along the same vertical line. For 27 species there can be a maximum of 13 phylogenetically independent contrasts, but since the phylogeny contains multiple-branching nodes the maximum here is only 12 (A–L). There are numerous ways of dividing the phylogeny into fewer contrasts, not shown. The numbers in the phylogeny, also from Sessions and Larson (1987), are times of divergence estimated from albumin immunological distances and Nei's electrophoretic distances on the assumption that these increase linearly with time (i.e. are good 'molecular clocks').

variances and covariances in other contrasts. The same is not true of random, even if non-overlapping, groups of species: the difference between a mouse and a macaque is not independent of the difference between a rat and a chimp. Figure 2 shows the data set of Sessions and Larson (1987) divided into 12 phylogenetically independent contrasts and Table 1 shows the

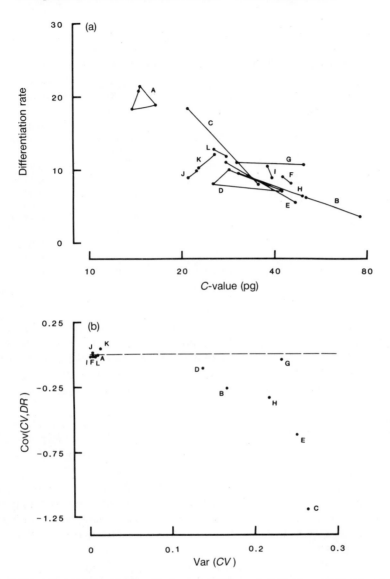

Fig. 3. Two views of the relationship between differentiation rate and *C*-value. (a) The standard species scatterplot shows a very highly significant negative correlation between species means ($r = -0.85$, $n = 27$, $p < 0.0001$). Lines connect members of phylogenetically independent contrasts, lettered as in Fig. 2. Note that the subfamily Desmognathinae (contrast A) is quite separate from the subfamily Plethodontinae (all the rest), indicating strong phylogenetic similarity. (b) Each contrast is reduced to a single point in this graph of the covariance of *C*-value and differentiation rate as a function of the variance in *C*-value. Note that the slope of the line connecting each point to the origin is the least-squares slope of the relation within contrasts. Data from Table 1.

Table 1

The variances of C-value (CV) and differentiation rate (DR), and their covariance calculated separately for each of the phylogenetically independent contrasts indicated in Fig. 2. Cov (X,Y) = [ΣXY − ΣXΣY/n]/(n − 1); n = number of species

Contrast	n	Var(CV)	Var(DR)	Cov(CV,DR)
A	4	0.00108	2.203	− 0.0032
B	2	0.01674	3.920	− 0.2562
C	2	0.02645	53.045	− 1.1845
D	3	0.01380	2.333	− 0.1050
E	2	0.02509	15.125	− 0.6160
F	2	0.00045	0.405	− 0.0135
G	2	0.02333	0.080	− 0.0432
H	2	0.02184	5.120	− 0.3344
I	2	0.00011	1.280	− 0.0120
J	2	0.00039	0.500	0.0140
K	2	0.00140	1.620	0.0477
L	2	0.00076	0.405	− 0.0176

variances and covariances of *C*-value and differentiation rate calculated for each contrast. A useful graphical method of representing the contrasts is to plot the covariance of *C*-value and differentiation rate as a function of the variance in *C*-value (Fig. 3b): the slope of the line connecting each data point to the origin is the least-squares slope of differentiation rate on *C*-value within that contrast [b = Cov(X,Y)/Var(X)].

To test for an association between *C*-value and differentiation rate one can simply apply a sign test to the covariance estimates. The null hypothesis is that the probability of a covariance being negative equals the probability that it is positive. Here 10 of 12 covariances are negative ($p = 0.039$, two-tailed), indicating a significant tendency for increases in *C*-value to be associated with decreases in differentiation rate, and vice versa, and rejecting the hypothesis that the species-level correlation is wholly due to phylogenetic similarity in these two characters. These results corroborate those of Sessions and Larson (1987).

As the force of these conclusions derives from the criterion of phylogenetic independence defined above, several comments about it seem appropriate. First, there are many different ways of dividing a phylogeny into independent contrasts and often there will be a trade-off between increasing the number of contrasts and increasing the proportion of variance which is 'within' contrasts (and thus being tested), as opposed to 'between' contrasts (and thus not being tested). It would seem appropriate to increase both statistics, but there is no obviously 'best' compromise. In the above analysis I maximized the number of contrasts, with the result that only 26 per cent of the variation in *C*-value and 27 per cent of the variation in differentiation rate lies within contrasts. In general for n species and a dichotomously branching phylogeny there are a maximum of $n/2$ phylogenetically independent contrasts (rounded down to the nearest integer), although fewer if the phylogeny contains multiple-branching nodes, as in the above example.

However the phylogeny is divided, recognizing independent contrasts by the above criterion will allow one to use more of the data than if one restricts contrasts to pairs of nearest relatives or to just one level of the Linnean hierarchy. Of course, this is only an improvement if the extra information used is correct: the effect of errors in the phylogeny will presumably be similar to taking random groups of species as independent contrasts, which in the limit is no improvement on the standard species-level analysis. Thus, when in doubt about the phylogeny, the conservative solution is to use multiple nodes.

On the other hand, the criterion of independent contrasts defined here is quite conservative in that values for a species are used only once; information on phylogenetic relationships among the contrasts is not used. As we have seen, this results in a maximum of $n/2$ degrees of freedom in the subsequent test of association. Is there an alternative? Felsenstein (1985) and Grafen (1989) make suggestions about how to combine data from different contrasts to construct yet more contrasts, thus deriving one degree of freedom for each higher node in the phylogeny (d.f. $= n - 1$ for a dichotomously branching phylogeny). However, these methods rely on particular assumptions about how information from different contrasts is to be combined. Felsenstein's (1985) method, one of those actually used by Sessions and Larson (1987), requires information on times of divergence for each node and the assumption that characters evolve by Brownian motion; Grafen (1989) suggests an even more complicated method which requires that a series of arbitrarily assigned branch lengths be correct. Although these methods may well be of some use, the method described here is more economical of assumptions and should be useful for this reason.

A good phylogeny does not exist for many taxa and comparative biologists studying these groups will have to extract contrasts from the Linnean classification. How might this be done? An obvious extrapolation of the above method involves using a variety of taxonomic levels, as follows: count each genus with two or more species as a contrast, remove them from the data set, count the tribes with two or more remaining genera as contrasts, remove them from the data set, count subfamilies with two or more remaining tribes as contrasts, remove them, and so on up the Linnean hierarchy. However, it should be recognized that this approach depends on each taxon being strictly monophyletic, at least with respect to the species in the data set. For better or worse, monophyly is not always a criterion in constructing higher taxa and so many Linnean taxa are not monophyletic. For example, the genus *Plethodon* in the salamander data set is paraphyletic (i.e. species on the same branch have been put in another genus, *Aneides*—see

Fig. 2), with the consequence that a contrast including the whole genus would not be independent of a contrast including both *Ensatina* and *Aneides*, two other genera in the tribe Plethodontini (Fig. 2). Thus, if contrasts are to be extracted from a Linnean classification and the taxa as represented in the data set are thought to be paraphyletic, then the conservative method is to use only one level of the Linnean hierarchy in defining contrasts. Five contrasts at the generic level or three at the tribal level can be identified in the salamander data set.

Tests of association between two qualitative variables may also be done using phylogenetically independent contrasts. Such variables include modes of sex determination, ovi- versus viviparity, and the sort of nucleotide found at a particular locus. Here I consider the simplest case of two variables with two states each and to do so have transformed the C-value and differentiation rate data into qualitative variables (Fig. 4). The exact same contrasts as used for the continuous data could be used, but there is a problem: to be useful in a test of association, a contrast must contain both states of both characters and only two of the original contrasts meet this additional criterion (C and E). Thus, unless the data set is very large, one will probably want to divide the data set in such a way that all contrasts contain both states of both characters. Unfortunately, the gain is slight in this data set, as there are a maximum of only three such contrasts (Fig. 4); even though the association in all three is in the same direction—large DNA with slow differentiation and vice versa—it is impossible to get a statistically significant result from a sign test. Thus, the method of independent contrasts can be easily extended to tests of association between qualitative characters, but in general one will need a much larger data set (or more detailed phylogeny) in order to get the same number of useful contrasts as for continuous variables. This approach provides an alternative to the methods of Ridley (1983, 1986; see Pagel and Harvey 1988a), which rely on the assumption that cladistic techniques such as outgroup comparison can accurately identify ancestral states. The test described here makes no assumptions about parsimony in evolution or the character states of ancestral species.

4. PHYLOGENETICALLY INDEPENDENT CONTRASTS AND EXPLICITLY HISTORICAL ANALYSES

One advantage of the above test of association is that it does not require data on the ages of each contrast; however, if such information is available then one can address a number of interesting historical questions. The most likely sort of data are molecular distances between extant species, which are being measured with increasing frequency for use in constructing phylogenies. The various uses of these data in aiding our understanding of phenotypic evolution has not, to my knowledge, been much considered, and is the topic

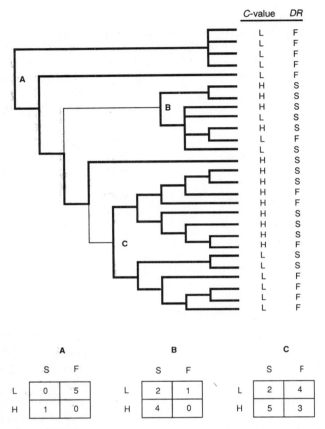

Fig. 4. Test of association between two qualitative variables. *Top*: the data of Fig. 2 have been transformed into qualitative variables by classifying *C*-values as higher or lower than the median (H, L) and differentiation rates as faster or slower than the median (F, S). The phylogeny is unchanged. Phylogenetically independent contrasts are again identified, with the proviso that each one contains both values for both variables. There are a maximum of three such contrasts in the above tree; one possible set of three, with approximately equal numbers of species in each is identified above (A–C). *Bottom*: each contrast is represented as a 2×2 contingency table of species values. Note that the direction of association in all three contrasts is the same.

of this section. I present a series of simple and emphatically exploratory analyses; again, all are illustrated with the salamander data, though, as will become obvious, significantly larger data sets will often be desirable. The branch lengths indicated in Fig. 2 are times of divergence (in millions of years) as estimated by a combination of albumin immunological distances and Nei's electrophoretic distances. These values may be regarded as a composite index of molecular divergence; they are available for nine independent pairwise contrasts with estimated times of divergence ranging from four to 33 million years ago (Table 2).

Table 2

Phenotypic and molecular divergence in nine pairwise contrasts. Molecular divergences are the sum of branch lengths from Fig. 2 and those for C-value and differentiation rate are the absolute values of the difference between the two species

Contrast	Divergence		
	Molecular	C-value	DR
B	20	0.183	2.8
E	33	0.224	5.5
F	14	0.030	0.9
G	23	0.216	0.4
H	27	0.209	3.2
I	12	0.015	1.6
J	4	0.028	1.0
K	18	0.053	1.8
L	5	0.039	0.9

4.1 Phenotypic and molecular divergence

One obvious point of departure is the relationship between molecular and phenotypic divergence (Fig. 5a, b). In these data it appears that differences in both C-value and differentiation rate are highly correlated with molecular divergence; indeed, the observed correlation coefficients (0.86 and 0.74 respectively) are about as high as those between the immunological and electrophoretic distances themselves ($r = 0.70$–0.88; Maxson and Maxson 1979). Thus, if molecular divergence is a good measure of the age of contrast then so is the divergence of C-value and differentiation rate, at least over the range considered here. The second point of interest is the intercept of regression lines fitted through the scatterplots. If speciation (cladogenesis) is typically associated with particularly rapid phenotypic evolution, as suggested by some models of evolution (e.g. Gould and Eldredge 1977), then the regression of phenotypic divergence on time should have a positive intercept. However, both intercepts are very close to the origin ($a_{CV} = -0.03$ and $a_{DR} = -0.12$), suggesting that no very large punctuation in either C-value or differentiation rate typically occurs at time of speciation (or, at least no larger a punctuation than may occur in molecular divergence).

4.2 Random walk models of evolution

Further analysis is suggested by considering various null models of evolutionary change. I have already referred to one such model, the random walks of Raup and Gould (1974; see also Raup 1977). In this model changes during a time period of unit length have zero mean and constant variance and

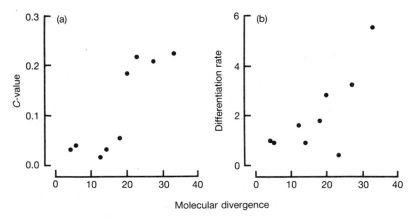

Fig. 5. Phenotypic and molecular divergence in nine phylogenetically independent contrasts. Arithmetic plots show strong positive correlations between molecular divergence and differences in both *C*-value (a) and differentiation rate (b). Note that both functions increase approximately linearly and that both intercepts are not significantly different from zero, the latter suggesting that there are no 'punctuations' in *C*-value or differentiation rate associated with speciation.

and changes in successive time intervals are independent; consequently it predicts that phenotypic variances—not differences—should increase linearly with time. If the model is altered such that changes in successive time intervals are positively correlated, as might be expected under persistent directional selection, then variances should increase faster than linearly with time; on the other hand, a negative correlation between successive changes would predict that variances should go up more slowly than linearly with time. Thus, if we write

$$\text{Phenotypic variance} = a(\text{Time})^b,$$

then the exponent b is a measure of the autocorrelation of changes in successive time intervals with $b=1$ when $r=0$. A log-log plot of the data, using the molecular divergence estimates of time, is shown in Fig. 6a. Unfortunately, the data are too few to give precise estimates: neither the individual slopes nor the common slope are significantly different from the random walk prediction of 1, but the confidence limits are wide and do not exclude slopes of 1/2 or 2 either. Similar analyses on much larger data sets would be of interest.

[It should be noted that this analysis is only approximate, as variances based on only two values, even log-transformed, are unlikely to be normally distributed. Furthermore, the least-squares slope calculated will tend to underestimate the true slope if there is a less than perfect correlation between

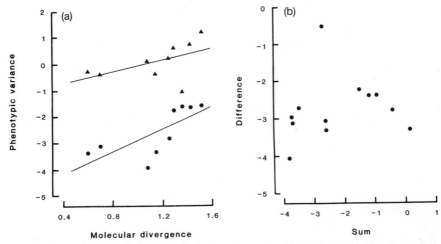

Fig. 6. Testing null models of evolution. The assumptions of the molecular clock and random walk model of phenotypic evolution together predict that phenotypic variance should increase linearly with molecular divergence. (a) Log-log plot of phenotypic variance as a function of molecular divergence for C-value (●) and differentiation rate (▲) in nine phylogenetically independent contrasts. Neither the individual slopes ($b_{CV}=2.1\pm1.79$ (95 per cent C.L.) and $b_{DR}=1.1\pm1.72$) nor the common slope ($b_C=1.6\pm1.12$) are significantly different from the predicted slope of 1. Data from Tables 1 and 2. (b) In the absence of molecular comparisons, or other data on times of divergence, one can test the null hypothesis that the variance in the log-transformed estimates of variance are equal for two characters by plotting the difference of the estimates [log Var(CV) $-$logVar(DR)] as a function of their sum [logVar(CV)$+$logVar(DR)] for 12 phylogenetically independent comparisons; a positive correlation would indicate that $\sigma^2_{\text{logVar}(CV)}>\sigma^2_{\text{logVar}(DR)}$ and a negative correlation the opposite. In fact, there is no significant correlation ($r=0.22$, $0.5>p>0.4$).

molecular divergence and time. Corrections for this bias are possible, but require information on the relative error variances associated with X and Y (Pagel and Harvey 1988*b*). Finally, the interpretations assume that rates of evolution have been constant through time: conceivably, an alternative explanation for a slope greater than 1 is that evolution, although remaining a random walk, has gradually slowed down over the time period being studied.]

A further test of the random walk model is to check whether rates of evolution have been homogeneous in different contrasts. Variances within each contrast may be standardized by dividing by the estimated age (molecular divergence) of the contrast and then compared using Bartlett's test (see Snedecor and Cochran 1980, p. 252 and tables in Pearson and Hartley 1966). Unfortunately, estimates of variance are very imprecise when based on only two values, and so this test has rather low power: not only are

there no significant differences between the standardized variances for either
C-value or differentiation rate, but there are also no detectable differences
between the unscaled variances (for which we know there are real differences,
since they are correlated with molecular divergence—Fig. 6a).

One conclusion that may be drawn from these results is that the simple
tests of the random walk described here, using independent contrasts, are
rather weak and will generally require many more than nine contrasts. One
possible alternative approach would be to use the molecular data to calculate
the complete series of contrasts defined by Felsenstein (1985). Such analyses
are more complex, for one has to recalculate the contrasts for each value of
the exponent b under test, but they do use more of the available information
and thus should provide more precise estimates. Unfortunately, using the 18
such contrasts extractable from this data set (see Sessions and Larson 1987)
one still cannot distinguish between exponents $b=1$ and $b=2$, nor detect
significant heterogeneity in the unscaled variances of either C-value or
differentiation rate. Thus, it seems the only solution is to increase the size of
the data set, if possible also increasing the range of contrast ages under study.

These tests of the random walk are obviously restricted to instances where
one has molecular data or some other measure of times of divergence; can
similar questions be addressed in the absence of such information? I suggest
that while one may not be able to test a specific evolutionary model, such as
the random walk, nevertheless one can test whether two phenotypic characters follow the same evolutionary model (whatever that may be). For
example, one might have a theory that naturally and sexually selected
characters should have different patterns of evolutionary change, or that
characters closely related to fitness should differ from more neutral characters. One approach would be to compare the partitioning of variation at
different taxonomic levels for the two characters. Unfortunately, formulae
for confidence limits of variance components in unbalanced data sets are not
known (Sokal and Rohlf 1981). Maximum likelihood methods of estimating
variance components would seem to offer a possible solution, and as such
would be worth further study; however, these are beyond the scope of this
chapter. Instead, I present an alternative method using independent contrasts, as follows. I suggested above that the slope of the regression of log-
variance on log-time across contrasts estimates the exponent in the relationship: Variance $= a(\text{Time})^b$. If two phenotypic characters have the same
exponent, then the variance among contrasts of the log-variances for one
character should equal that for the other (i.e. H_0: $\sigma^2_{\text{logVar(CV)}} = \sigma^2_{\text{logVar(DR)}}$). One
tests this hypothesis by correlating the sum of the log-variances for the two
characters and their difference, across contrasts; a significant correlation
indicates a significant difference in the two variances (Snedecor and Cochran
1980, pp. 190–1). Applying this test to the salamander data reveals no
significant difference between the variance among contrasts for C-value and
differentiation rate (Fig. 6b), and so we cannot reject the hypothesis that
these two characters have the same exponent. Two points to note: like the

previous test, this one is only approximate as log-variances are unlikely to be normally distributed (although the situation may be improved slightly as they will often be based on more than two values), and significant differences in variances may be due not only to differences in exponent (i.e. the variance due to time in Fig. 6a), but also, perhaps, to differences between characters in the heterogeneity of rates of evolution (i.e. the residual variation).

4.3 Phylogenetic inertia

Returning to the uses of data from molecular comparisons, or other information on times of divergence, one can ask whether the significant negative association between C-value and differentiation rate changes as a function of the age of the contrast. Such a change might be predicted by an hypothesis of phylogenetic inertia. This process occurs when changes in X result in changes in selection pressure on Y, but there has not been enough time for Y to respond completely, such that the population mean is lagging behind the optimum; the faster X changes, the more Y should be lagging. Thus, an hypothesis of phylogenetic inertia would predict that the slope of the relation within contrasts should become shallower as one considers more recent contrasts. Of course, it is possible that differentiation rate is not responding directly to changes in C-value, but that they are each responding—in opposite directions—to changes in some other unknown variable Z; if so, then an association between the slope of differentiation rate on C-value within contrasts and the age of the contrast would suggest that they differ in their relative rates of response to Z—one is lagging more than the other. As can be seen from Fig. 7, there is in fact no significant association. This result should not come as a great surprise: first, the mechanism relating C-value to differentiation rate is not thought to be evolutionary—increases in C-value do not *select* for decreases in differentiation rate—but rather biochemical, and thus immediate. Secondly, even for two variables related through natural selection, the selection coefficients or heritabilities would have to be quite low to show lags in contrasts over four million years old.

An analogous test can be devised in the absence of molecular data, though it will probably be weaker. Using a phylogeny one can often tell which of two contrasts is more recent, and then compare the slopes of those contrasts. There are five such pairs of contrasts to be made in the salamander phylogeny and four of them go against the hypothesis (Table 3).

These tests for phylogenetic inertia are very similar to tests for consistent changes in slope with taxonomic level (i.e. a consistent increase or decrease in slope as one calculates slopes for species within genera, genera within families, families within orders, etc. (e.g. Harvey and Mace 1982)). However, they are not to be confused with various tests of taxonomic similarity proposed by several other authors. Cheverud *et al.* (1985) estimate 'phylo-

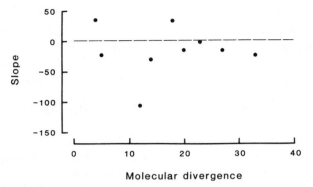

Fig. 7. Test for phylogenetic inertia. As there is a negative correlation between differentiation rate and *C*-value, an hypothesis of phylogenetic inertia predicts a negative correlation between the slope of the relation within contrasts and the age of the contrast (i.e. older contrasts should have more negative slopes). Here, age is estimated by molecular divergence and there is no indication of a trend: the slope of the association does not depend on the age of the contrast. Data from Tables 1 and 2.

genetic inertia' with an autocorrelation coefficient and Derrickson and Ricklefs (1988) talk about phylogenetic constraints in terms of the slope and shape of a relationship differing in different taxa. However, these are measures of taxonomic similarity, akin to a nested ANOVA; as such they may (or

Table 3
Sign test for phylogenetic inertia when data on times of divergence are not available

Predicted	Observed?
C < B	Yes
D < E	No
G < F	No
H < I	No
K < L	No

The null hypothesis is that the relative branching order of the contrasts is independent of the relative value of the slopes within contrasts; the hypothesis of phylogenetic inertia predicts that older contrasts will have more negative slopes. Thus, for example, we know from the phylogeny that contrast C is older than B, and so we predict that it should have a more negative slope, as indeed it does (-44.8 vs. -15.3). However, contrast D is older than E, yet has a higher slope (-13.0 vs. -24.5).

may not) be useful (see also Bell, 1989). The test described here is specifically for a trend in the slope of the relation with time. Although phylogenetic inertia is one possible cause of taxonomic similarity, it is by no means the only, and the two concepts should not be synonymized.

5. OTHER USES FOR INDEPENDENT CONTRASTS

The method of independent contrasts is useful in comparative biology to deal with a specific problem: one wants to test a causal hypothesis relating X and Y, but there is a plausible process with no causal link between X and Y which nevertheless will routinely generate significant correlations between them. In interspecific comparisons this process is the sporadic branching of the evolutionary tree and the production of descendant phenotypes by modification of ancestral phenotypes. Similar statistical problems may be widespread. For example, correlations among individuals are very common in biology, yet close relatives are likely to be more similar than distant relatives. Those who study lakes regularly regress limnological variables on each other using lakes as data points, yet closely neighbouring lakes are presumably more similar than those far apart. In both cases, one approach to dealing with this underlying structure in the data is to use independent contrasts. Does the relationship hold within families? Within regions?

6. SUMMARY

Interspecific correlation is a popular means of testing evolutionary and ecological theory. However, computer simulations by Raup and Gould (1974) demonstrate that statistically significant correlations can easily arise between two variables undergoing independent random evolutionary walks. Indeed, significant associations are likely whenever closely related species are more similar than distantly related species, for then species with similar X-values are likely to be closely related, and thus have similar Y-values, independent of any causal mechanism relating X and Y. Here, I discuss an alternative test of association which avoids this problem and thus can provide a more powerful test of specific causal hypotheses than simple interspecific correlations.

The method involves two steps: first, dividing the data set into a series of independent replicate contrasts, and secondly, testing for a consistent association between changes in X and changes in Y. Several previous discussions of this method have defined the independent contrasts either as pairs of closest relatives or as taxa at one level of the Linnean hierarchy (e.g. genera); here I suggest a novel method of recognizing independent contrasts in a known phylogeny. If contrasts are represented as a path joining the member species through the phylogenetic tree, then contrasts are independent if their paths do not meet at any point. For n species there are a maximum of $n/2$ phylogenetically independent contrasts, given a dichotomously branching phylogeny; multiple-branch points can be easily accommodated, though they reduce the maximum number of contrasts. Complications arise if contrasts must be extracted from a Linnean classification: if the taxa as represented in the data set are thought to be monophyletic then exactly analogous methods apply; however, if paraphyletic, then the conservative solution is to use contrasts from only one level of the Linnean hierarchy. Each phylogenetically independent contrast contributes one degree of freedom to the subsequent test of association; any association which is significant by this test cannot be due to phylogenetic similarity. A sign test on the covariance of characters within contrasts provides a conservative nonparametric test. The method can be easily extended to test for an association between two qualitative variables, although in general one will need a larger data set in order to get the same number of contrasts.

Although this test of association is purely ahistorical in nature and does not require data on times of divergence, such information, if available, can be used together with the phylogenetically independent contrasts in explicitly historical analyses. The relationship between phenotypic and molecular divergence can be used to look for evidence of 'punctuations' associated with speciation and to test null models of evolutionary change (e.g. random walks). Further, if two phenotypic characters are found to be correlated, the relationship between the slope of the association within contrasts and the age of the contrast can be used to test hypotheses of phylogenetic inertia. Similar analyses are also presented for the much more common instance of there not being any data on times of divergence other than the phylogeny, although the tests are correspondingly weaker.

All analyses are illustrated using Sessions and Larson's (1987) data on C-value and differentiation rate in plethodontid salamanders. The main conclusions are:

1. Increases in C-value tend to be associated with decreases in differentiation rate, and vice versa, across 12 phylogenetically independent contrasts.

2. There is no indication of punctuations in either C-value or differentiation rate at time of speciation.

3. The evolution of C-value and differentiation rate is compatible with a null model of random walk; however, the tests have rather low power and more data will be necessary to provide a more precise test.

4. There is no evidence of phylogenetic inertia in the relationship between C-value and differentiation rate.

Throughout the chapter these methods are compared to various proposed alternatives. I conclude with several brief comments on the application of the method of independent contrasts beyond the problem of species comparisons.

ACKNOWLEDGEMENTS

I have had many useful discussions on comparative methods with Graham Bell, Gilbert Cabana, Julee Greenough, Paul Harvey, Vasso Koufopanou, Andrew Read, and, in particular, Alan Grafen. Joseph Felsenstein, Alan Grafen, Paul Harvey, Vasso Koufopanou, Rob McLaughlin, Mark Pagel, Mike Paine, and Linda Partridge made useful comments on a previous draft.

REFERENCES

Bell, G. (1989). A comparative method. *Amer. Natur.* **133**, 553–71.

Brown H. S. (1961). Differential chiasma frequencies in self-pollinating and cross-pollinating species of the genus *Gilia. El Aliso* **5**, 67–81.

Cheverud, J. M., Dow, M. M., and Leutenegger, W. (1985). The quantitative assessment of phylogenetic constraints in comparative analyses: sexual dimorphism in body weight among primates. *Evolution* **39**, 1335–51.

Clutton-Brock, T. H. and Harvey, P. H. (1984). Comparative approaches to investigating adaptations. In *Behavioural ecology: an evolutionary approach*, (2nd edn) (ed. J. R. Krebs and N. B. Davies), pp. 7–29. Sinauer Associates, Sunderland.

Derrickson, E. M. and Ricklefs, R. E. (1988). Taxon-dependent diversification of life-history traits and the perception of phylogenetic constraints. *Funct. Ecol.* **2**, 417–23.

Felsenstein, J. (1985). Phylogenies and the comparative method. *Amer. Natur.* **125**, 1–15.

—— (1988). Phylogenies and quantitative characters. *Ann. Rev. Ecol. Syst.* **19**, 445–71.

Gould, S. J. and Eldredge, N. (1977) Punctuated equilibria: the tempo and mode of evolution reconsidered. *Paleobiology* **3**, 115–51.

Grafen, A. (1989). The phylogenetic regression. *Phil. Trans. Roy. Soc.* (in press).

Harcourt, A. H., Harvey, P. H., Larson, S. G., and Short, R. V. (1981). Testis weight, body weight and breeding system in primates. *Nature* **293**, 55–7.

Harvey, P. H. and Clutton-Brock, T. H. (1985). Life history variation in primates. *Evolution* **39**, 559–81.

—— and Mace, G. M. (1982). Comparisons between taxa and adaptive trends: problems of methodology. In *Current problems in sociobiology*, (ed. King's College Sociobiology Group), pp. 343–61. Cambridge University Press.

Maxson, L. R. and Maxson, R. D. (1979). Comparative albumin and biochemical evolution in plethodontid salamanders. *Evolution* **33**, 1057–62.

Nevo, E., Beiles A. and Ben-Shlomo, R. (1984). The evolutionary significance of genetic diversity: ecological, demographic and life history correlates. In *Evolution-

ary dynamics of genetic diversity, Lecture Notes in Biomathematics No. 52, (ed. G. S. Mani), pp. 13–213. Springer, Berlin.

Pagel, M. D and Harvey, P. H. (1988*a*). Recent developments in the analysis of comparative data. *Q. Rev. Biol.* **63,** 413–40.

—— and —— (1988*b*). The taxon-level problem in the evolution of mammalian brain size: facts and artifacts. *Amer. Natur.* **132,** 344–59.

Pearson, E. S. and Hartley, H. O. (1966). *Biometrika tables for statisticians*, Vol. 1, (3rd edn). Cambridge University Press.

Raup, D. M. (1977). Stochastic models in evolutionary paleontology. In *Patterns of evolution as illustrated by the fossil record* (ed. A. Hallam), pp. 59–78. Elsevier, Amsterdam.

—— and Gould, S. J. (1974). Stochastic simulation and evolution of morphology— towards a nomothetic paleontology. *Syst. Zool.* **23,** 305–22.

Read, A. F. (1987). Comparative evidence supports the Hamilton and Zuk hypothesis on parasites and sexual selection. *Nature* **328,** 68–70.

Ridley, M. (1983). *The explanation of organic diversity: the comparative method and adaptations for mating*. Clarendon Press, Oxford.

—— (1986). The number of males in a primate troop. *Anim. Behav.* **34,** 1848–58.

Sessions, S. K. and Larson, A. (1987). Developmental correlates of genome size in plethodontid salamanders and their implications for genome evolution. *Evolution* **41,** 1239–51.

Snedecor, G. W. and Cochran, W. G. (1980). *Statistical methods*, (7th edn). Iowa State University Press, Ames.

Sokal, R. R. and Rohlf, F. J. (1981). *Biometry*, (2nd edn). W. H. Freeman, New York.

Stearns, S. C. (1983). The influence of size and phylogeny on patterns of covariation among life-history traits in the mammals. *Oikos* **41,** 173–87.

Strong, D. R., Lawton, J. H., and Southwood, R. (1984). *Insects on plants: community patterns and mechanisms*. Blackwell, Oxford.

Wootton, J. T. (1987). The effects of body mass, phylogeny, habitat, and trophic level on mammalian age at first reproduction. *Evolution* **41,** 732–49.

Caste and change in social insects

DEBORAH M. GORDON

Most models of the organization of social insect colonies start with a collection of individuals, each of which is assigned to a particular task. The notion of caste is fundamental to current thinking about social insects. The term was originally used to distinguish classes of insects within a colony, the members of which show strong differences of morphology and reproductive status, such as queens, workers, and males. In the mid-1960s, 'caste' came to refer to a finer subdivision: classes of workers that do particular tasks (Wilson 1963). Species with 'physical castes', such as some polymorphic ant species, contain workers of various different sizes, and those of a particular size class do a particular task. In species with 'temporal castes', such as honeybees, and many ant species, task is age-dependent; a class of adult workers of a particular age tends to do a particular task. The use of the word 'caste', which also refers to stable distinctions between groups in human societies, reinforces the notion of a task as determining a fixed attribute of particular, individual insects.

About 20 years ago, results on task specificity of individual workers began to suggest a working hypothesis for the study of the evolution of colony organization (Wilson 1968): The behaviour of a colony is a result of the numerical distribution of workers into a set of essential tasks. For example, a colony with more individuals of the caste that forages will forage more; a colony with more individuals devoted to the task of brood care will raise more brood, or raise them faster. Based on this view, Wilson (1968) introduced the notion of a caste distribution, that is, the proportion of colony members belonging to each caste, as a species character that may evolve towards the most adaptive one. For example, if natural selection favours colonies that forage more, a species may evolve colonies that contain a larger proportion of individuals of the foraging caste.

This line of argument underlies the work of Oster and Wilson (1978), to date the most comprehensive and influential attempt to model the organization and evolution of social insect colonies. In their words (p. 21):

The guiding proposition of our inquiry is that variations in caste structure and division of labor reflect differing adaptations on the part of individual species of social insects. Thus, caste is not just central to social organization; it should provide the key to the ecology of social insects, insofar as those insects differ from their solitary counterparts. We regard many of the principal processes of colony life ... as

56   Deborah M. Gordon

subordinate to the evolution of caste. We postulate them to be the enabling devices by which labor is allocated and by which the colony as a whole precisely adjusts its relationship to the nest environs.

Here, I argue that this approach to the study of colony organization is now being revised in the light of recent empirical results. I will concentrate on the literature on ant behaviour, with some reference to honeybee behaviour. First I outline the evidence that colony behaviour does not depend on caste distributions, and then discuss the evolutionary implications of this conclusion. (I use 'colony behaviour' to refer to a sequence of transient behavioural states. These could be specified in various ways. For example, one could list the activities of each colony member, or use some measure, such as numbers or proportions of workers, of the intensity of colony effort devoted to certain activities). What matters in natural selection is what a colony does, for example, how much it forages or how many reproductives it produces in particular conditions. I will argue that what a colony does, its behavioural phenotype, cannot be predicted reliably by specifying the habitual task of each of its workers and counting the numbers of workers in each task category, or caste. There are three reasons why colony behaviour is not a simple function of caste distributions: (1) individuals switch tasks; (2) the intensity of colony effort devoted to various tasks changes; and (3) different colonies vary in the dynamics that link colony behaviour and environmental conditions.

1. COLONY BEHAVIOUR DOES NOT DEPEND SIMPLY ON CASTE DISTRIBUTIONS

1.1 Individuals switch tasks

There is a substantial literature to show that individuals tend to do particular tasks. A relatively new line of research examines the relation between individual task and colony environment. It is becoming increasingly clear that when colony conditions change, individuals switch tasks. Both task specificity and switching have been reviewed in Calabi (1987), who lists 26 studies showing task switching among age classes and 27 showing task-switching among size classes. Here I cite a few examples. Task-switching has been shown to occur in several kinds of situations.

1. *Disturbances* that obstruct colony activities, temperatures that endanger the brood, and newly available food sources (Meudec and Lenoir 1982; Mirenda and Vinson 1981; Gordon, in press), can all cause individuals to change tasks to meet new exigencies of the colony environment.

2. *Changes in numbers of workers.* When workers engaged in one task are removed from the colony, workers previously committed to other tasks will fill in for the missing workers (Wilson 1984). Carlin and Holldobler (1983)

added pupae of different species of the carpenter ant *Camponotus* to existing colonies, and found that differences in the species composition of a colony caused workers to change tasks.

3. *Changes in the age structure of the colony*. In species in which age cohorts do particular tasks, removals of older age cohorts accelerate the usual sequence of age polyethism for younger workers (in ants, McDonald and Topoff 1985; in bees, Winston and Punnett 1982; Winston and Fergusson 1985).

The question of task-switching arises in any attempt to specify the relation between the set of available individuals and the set of necessary tasks. There might be a simple one-to-one relation: there are x tasks, x individuals, and each individual always does one, unique task. There might be a random relation: any individual is equally likely to do any task. There might be x groups of individuals, and x tasks, each group consisting of individuals that always do a particular task. This is the relation implied by the simplest view of the caste concept. In fact, most evidence points to a fourth relation. There are distinct groups of individuals, which are sometimes classifiable by size or age. At any time, each group contains individuals that usually do a particular task. However, when environmental conditions and colony composition change, new groups form. That is, when conditions change, individuals switch from one task group to another.

It may seem intuitively obvious that an organization that utilizes a division of labour based on permanently specialized individuals is inherently more efficient. But is it? Is an organization consisting of specialized individuals necessarily more efficient than one in which each individual is capable of a variety of tasks? This question suggests an analogy with human industry, from which the phrase 'division of labour' is borrowed. When Adam Smith introduced the idea of division of labour in 1776 in his book *The wealth of nations*, he suggested three reasons why it constitutes an efficient way to organize manufacture: (1) Workers will learn to do their jobs better if they repeat the same job over and over. (2) Workers have an opportunity to slack off when they change from one job to another. 'The habit of sauntering and of indolent careless application, which is ... necessarily acquired by every country workman who is obliged to change his work and his tools every half hour ... renders him almost always slothful and lazy, and incapable of any vigorous application even on the most pressing occasions.' (3) When a man has only one task to do, he is likely to occupy his mind with thinking up a machine that can do it better.

Since Adam Smith's time, there has been considerable controversy about whether division of labour in a factory increases its productivity (Rose 1988). Advocates of assembly line methods disagree with those, such as the manufacturers of Saab automobiles, who believe that people work more efficiently when they have a variety of jobs to do.

However, none of Adam Smith's three advantages of individual specialization seem to apply directly to social insects. Insects are not likely to invent machines, and there is no evidence that individual specialization in animals promotes the evolution of tool use. Workers do show considerable variation in activity level, but there is no evidence that spells of worker inactivity are especially likely to occur if workers change from one task to another, rather than performing bouts of a single task, more or less interrupted by bouts of inactivity. Division of labour does not seem to reduce 'indolent careless application'. Most discussion of the division of labour in social insects seems to centre around the first of Adam Smith's reasons for it. It is not generally clear whether workers learn to do their tasks better by repeating them (although see Heinrich 1979), and experimental studies have revealed little capacity for learning an artificial task in social insects (e.g. for ants, Weiss and Schneirla 1976; Bernstein and Bernstein 1969).

However, workers may be physiologically adapted to particular tasks, and because all social insect workers are fully mature adults, one may expect such adaptations to be permanent. This is why there has been considerable attention to the ways that a worker's size may contribute to the efficiency with which it performs a particular task (Wilson 1968). But most ant species are monomorphic, that is, have only one size of worker; only 44 out of 263 genera contain polymorphic species (Oster and Wilson 1978). Task-switching goes on even in polymorphic species (Wilson 1984). Proliferating examples of task-switching suggest that the 'ergonomics of caste', that is, the fit between an individual's body size and its task, may not be the primary selective force shaping the evolution of social insect behaviour.

Oster and Wilson's models include predictions about the evolutionary consequences of behavioural flexibility. In their formulation, there is a range of tasks that an individual can perform successfully, and the magnitude of this range may vary. Thus tasks are assigned to individuals, and the repertoire of each individual may be more or less broad. Their prediction is that in species in which individuals are more flexible, that is, can perform a wider range of tasks, fewer distinct castes are needed. Their concept of behavioural flexibility maintains the primacy of caste, and considers its elasticity: within a behavioural caste, how much can a specialized individual change its task?

Oster and Wilson (1978) argue that the behavioural repertoires of social insects have decreased in the course of evolution; that is, in more advanced hymenopterans, such as ants, each individual is capable of a smaller number of different activities than in more primitive species such as solitary wasps. They suggest that one reason for this may be the increased efficiency of specialization. But the original assertion, that individual specialization has increased in colonial species in the course of evolution, is open to question. The assertion is based on estimates of the sizes of behavioural repertoires. Recently, it has been pointed out that these estimates, and comparisons among species of such repertoires, are problematic at best (e.g. Jaisson et al.

1988). To estimate the behavioural repertoire of a species, an observer must classify all acts into different types. The number of types is the magnitude of the reportory, and different observers tend to make very different classifications.

Oster and Wilson's models may seem to demonstrate that the social organization of colonies must depend on individual specialization because this is the most efficient type of organization. This is not the case. The models provide a theoretical framework within which to test the hypothesis about the efficiency of individual specialization. An unexpected result of such investigations is that individual specialization is not as pervasive as previously believed. Individuals change tasks. Arguments for the efficiency of individual specialization in social insects were proposed to explain the apparent pervasiveness of specialization. Put very simply, the argument is 'If individual specialization is more efficient, more specialized individuals will evolve'. If this assertion is true, any absence of permanent individual specialization is evidence against its efficiency.

Oster and Wilson (1978) do argue that parallel operations are more efficient than operations carried out in series. They explain this very clearly, using an example (pp. 13–14):

Suppose that several persons were trying to stack a specified number of bottles one on top of the other in the middle of a room. If one person tried it from the floor up, and then another started fresh, and so on, the group result would equal the average individual result. Now suppose that two or more persons work together. If their competence is low, they will tend to cancel out one another's actions. The chance of making the right combination of correct movements will be even lower than the chance of doing one thing correctly by itself, and fewer bottles will be stacked. However, as individual competence increases, the group will reach a point at which they are able to put together the right combination of moves and balance a significantly higher number of bottles. This improvement will be greater still if they divide the labor. For example, one can hold and readjust the bottles already stacked, a second can put more bottles on top, while a third can fetch still more bottles to add to the attempt.

The first part of the example describes an operation carried out in series; each person attempts to complete a series of tasks. A group of individuals can also work in parallel, each contributing to the stacking operation by doing one of the tasks involved. If individuals are competent enough, a parallel operation will be more efficient than a series one.

But the increased efficiency of parallel operations is not necessarily due to individual specialization. In the paragraph quoted above, 'divide the labor' does not imply consistent individual specialization, and this ambiguity may have misled some readers. If a particular individual holds the bottles one time and fetches more bottles the next, the advantages of working in parallel will not be diminished (Jeanne 1986a,b). Individual specialization is only an advantage if one individual is especially good at holding bottles, and another at fetching them, and so on. Oster and Wilson's examples of the advantages of individual specialization in social insects are based on polymorphic species. In cases where morphological differences make particular tasks easier for certain individuals, the advantages of individual specialization are clear. But as we have seen, such cases are relatively rare.

In other parts of Oster and Wilson's book, the advantages of consistent individual specialization are confounded with advantages due to replication, or the presence of multiple workers able to do a particular task. Comparisons are frequently made between the efficiency of a solitary wasp, that does a range of tasks herself, and that of a colony of specialized individuals. But there are several differences between these two organizations. One difference is that a group of workers can be in more places, and do more things, per unit time, than a single worker can. This leads to economies of scale: the benefits in productivity of an increased worker force outweigh the average costs of supporting it. In the discussion following the paragraph quoted above, Oster and Wilson (1978) point out that the reliability of a system is increased by redundancy. As they put it (p. 15) 'if a designer is given two sets of "parts", it is better to build a single system with redundant components than to build two separate systems'. If one worker does not accomplish a task, another worker might. Economies of scale, and enhanced reliability due to replication, do not depend on permanent individual specialization. They are advantages that arise from the presence of more workers. These advantages remain even if individual workers do not specialize on a particular task.

Individual specialization, to the extent that it exists in any species, may contribute to the efficiency of that species' social organization. But, as Oster and Wilson point out, the evolution of behaviour may be strongly affected by other factors. Until we have investigated the evolutionary consequences of other aspects of colony organization, there are no grounds for assuming that natural selection acts primarily to promote individual specialization. It may be, for example, that the efficiency of specialized individuals is less important than the colony-level efficiency of switching rules. A colony's ability to re-allocate workers as colony needs change may be more important than the efficiency of individuals at particular tasks. An individual would perform one task one day, another the next. In this case, the notion of caste, which posits a fixed relation between individual and task, is difficult to apply.

The problem of individual specialization raises methodological questions as well as theoretical ones. If individuals change tasks frequently, it may not be appropriate to describe or model the system in terms of what individuals do. This would involve specifying both individuals and conditions: Ant 1 does task x in condition A, task y in condition B, task z in condition C. Ant 2 does task y in condition A, task z in condition B, and so on. From the perspective of evolutionary ecology, as Oster and Wilson (1978) emphasize, the behavioural phenotype of the colony is its important characteristic. What matters is how colonies vary in the extent to which task x, y, and z are done.

If individuals frequently change tasks, it may be sufficient to specify the relationship between numbers of ants and various conditions, without regard to individual identity.

Individual task fidelity and the variability of environmental conditions may be linked. In many species, colonies contain large numbers of reserve workers. Presumably these reserves contribute to the flexibility of colony behaviour by providing individuals that can be channelled into tasks when they are needed. The dynamics of task switching are the rules that relate individuals and tasks in various environmental conditions. These rules govern both the recruitment of reserves to do particular tasks, and individual decisions to change tasks in particular circumstances. Selection may act as strongly on the dynamics governing individual decisions to change tasks in particular circumstances, as it does on the relation between an individual's body size and its usual or baseline task.

For example, the dynamics of task switching in harvester ant colonies reflect the species' ecological situation in a community in which a diverse guild of granivorous species are all competing for food (Davidson 1980). A series of perturbation experiments with marked individuals shows a clear trend in the direction of task-switching (Gordon, in press). In undisturbed colonies, various exterior tasks are each performed by a distinct group of workers. But when a new source of seeds suddenly becomes available, workers switch tasks. In this situation, workers previously committed to any other task, or to the reserves available for any other task, will switch tasks to forage. In contrast, no other group of workers will change tasks to do nest maintenance work. Foraging acts as a sink; nest maintenance is a source. It appears that the high priority of foraging may have shaped this species' rules for changes of task. To understand the dynamics that determine how and when individuals change task, it will be essential to compare such dynamics in a variety of ecological communities.

1.2   The intensities of colony activities change

Within a colony, individual workers show considerable variation in the extent to which they are active (Plowright and Plowright 1988; Lindauer 1961; Herbers 1983). Some individuals tend to be consistently more active than others; others show extremely variable activity levels. This means that even taking the tasks of particular individuals to be fixed, the numbers of workers that actually engage in them are not necessarily constant. More foragers does not necessarily imply more foraging.

The numbers that actually engage in a task may change according to changes in colony requirements for the products of that task (e.g. Kolmes 1985; Moore et al. 1987; Sorensen et al. 1985). For example, the work of Seeley (1986) and Schmid-Hempel and Schmid-Hempel (1987) shows that the activity level of individual honeybee foragers on a given day depends on the amount of nectar obtained in the colony that day. The amount of nectar obtained is determined not just by numbers of foragers but by the dynamics relating foraging behaviour and nectar stores.

The numbers that actually engage in one task may also vary according to changes in the behaviour of workers engaged in other tasks. For example, I found that in harvester ants, nest maintenance work and foraging are done by two distinct groups of workers, and foragers do not switch tasks to do nest maintenance work (Gordon, in press). But when experimental perturbations cause larger numbers of workers to do nest maintenance work, the foragers become less active. Several exterior activities, including patrolling, are interdependent in this way (Gordon 1986, 1987).

Perturbations that change the numbers engaged in one task, cause changes in the numbers engaged in other tasks. When two or more of these perturbations are done simultaneously, the results are non-additive. The response to combined perturbations is different from the sum of responses to single ones. This means that workers engaged in one task do not have a simple, all-or-nothing response to changes in other activities. Instead, the response is modulated according to the ways that all other activities are affected.

Calabi (1987) makes a strong case that the notion of 'caste' is obsolete, on the grounds that there is so much evidence that individuals switch tasks. She goes on to present a model of colony organization in which task-switching functions to keep the colony's labour profile constant. In other words, she assumes that colonies maintain stable numerical distributions of active workers into various tasks, and that individuals switch tasks to keep these distributions at constant levels. The results described above show that this assumption is unrealistic. Colonies alter their labour profiles significantly in complex but rule-governed ways. Task-switching does not function merely to maintain constant labour profiles. It also contributes to changes in numbers of workers engaged in different tasks.

The result that colonies change their allocations of efforts to various tasks, depending on environmental conditions, has important implications for future research. One is that activities cannot be understood independently of each other. For example, models of foraging in ants (e.g. Taylor 1978) predict how forager behaviour and morphology will evolve in response to conditions affecting the efficiency of foragers. However, forager behaviour changes significantly in response to events affecting workers engaged in other tasks. This means that selective pressures on foraging success are as likely to act on the dynamics of interactions between foragers and other worker groups, as they are on the characteristics of foraging itself.

Several time-scales, from the hourly to the ontogenetic, are relevant to the dynamics of colony behaviour. Individuals adjust their behaviour based on moment-to-moment (Meudec and Lenoir 1982) and hour-to-hour (Seeley 1986; Gordon 1987) changes of colony environment. The dynamics of colony behaviour also change as a colony matures. In younger colonies, individuals appear to be more flexible with respect to task (Lenoir 1979; Gordon, in press). I found that in 2-year-old harvester ant colonies, interactions among worker groups are different from those in 5-year-old colonies. That is, events affecting workers engaged in one task elicit changes in the numbers engaged in other tasks, and such changes are qualitatively different in older and younger colonies. For example, when older colonies are exposed to several simultaneous disturbances, their behaviour emphasizes foraging. That is, as things get worse for an older colony, it devotes more effort to foraging. This is not the case in younger colonies. Also, different groups of older colonies show similar responses to the same experiments. But the responses of different groups of younger colonies to similar experiments are much less consistent than those of older colonies. As ants live only for a year, behavioural differences in 2- and 5-year-old colonies cannot be due to the accumulated experience of 5-year-old ants. Instead, the organization of the colony must change as it matures. The dynamics of interactions among worker groups, which determine the allocation of colony effort, depend on colony age.

Caste distributions alone, or ontogenetic changes in a colony's caste distributions, are insufficient to predict these results. Specifying the task that individuals usually do, even if they do not switch tasks, does not specify how much of the task will be done in particular conditions. That is, caste distributions do not predict allocation of effort. To predict colony behaviour, we also need to know how the allocation of effort varies as the environment changes.

1.3   Variation among colonies

No one expects all colonies of a species to behave in exactly the same way. But in the past 20 years of research on social insects, two trends seem to reinforce each other: the use of caste distributions to explain colony behaviour, and concern with variation in behaviour among individuals within a colony, rather than with variation among colonies.

The literature on task-specificity in ants provides an example. To determine how likely individuals are to continue with particular tasks, it is necessary to mark individuals engaged in particular tasks, and to observe what tasks they do subsequently. I surveyed the literature of the past 10 years for reports of such experiments. Table 1 shows some features of the experimental design of such studies. I exclude references to work that does not examine behaviour independently of physical caste, that is, which assumes but does not document a correlation between behaviour and worker size. Most of the studies are based on data from a very small number of colonies. Data for different colonies are frequently pooled, making it

Table 1

Variation among colonies in investigations of task fidelity. F = field, L = laboratory
X = discussed, – = not discussed

Lab. or field	No. of colonies observed	Data given for colonies		Variation		Reference
		Pooled	Separately	Among colonies	Within colonies	
F	3	?		−	−	Calabi *et al.* (1983)
L	1	1		−	−	Carlin (1981)
F	4	4		−	−	Fowler and Roberts (1980)
L	5		5	−	−	Gordon (1984)
F	44		44	X	−	Gordon (in press)
F	21	19		X	X	Herbers (1979)
L	3	3		X		Herbers (1983)
F	1	1		−	−	Herbers (1977)
L	1	1		−	−	Hölldobler and Wilson (1986)
L	5	5		−	−	Lenoir and Ataya (1983)
L	2	2		−	−	Meudec and Lenoir (1982)
L	3	?		−	−	Mirenda and Vinson (1981
F	4	4	2	−	X	Porter and Jorgensen (1981)
F	3	3	2	X	X	Rosengren (1977)
L	1	1		−	−	Traniello (1978)
L	1	1		−	−	Wilson (1976)
L	1	1		−	−	Wilson (1978)
L	1	1		−	−	Wilson (1980)
L	1 per species	1?		−	X	Wilson (1984)
L	1	1		−	−	Wilson and Hölldobler (1985)
L	3	3		−	−	Wilson and Hölldobler (1986)

impossible to discern variation among colonies. Fifteen out of 21 studies are done in laboratory conditions.

The way least likely to reveal task-switching is to look at small numbers of colonies in the laboratory. Small numbers of colonies would tend to vary less in colony requirements than larger numbers because a small sample can capture less of the natural range of variation than a large one. Laboratory conditions tend to be more stable than field conditions, so changes in colony environment that lead to task switching would be less likely to occur in the laboratory.

A methodology that ignores variation has wide-ranging consequences for theories of social organization in insect colonies. Consider an example from a different field. A developing embryo passes through critical periods during which particular stimuli can drastically alter the course of differentiation. An investigator who observes but one embryo, in conditions under which these stimuli do not occur, would never guess at the existence of these dynamics. An investigator who observes a large number of embryos may happen to observe one that undergoes alteration, and decide to examine the causes of the alteration. Finally, an investigator who undertakes a series of experiments in which embryos are exposed to crucial stimuli at different stages of development, will be most likely to understand the dynamics of differentiation.

Observations of few colonies, in very stable conditions, over short time periods, are likely to lead to uniform results, which can be deceptive in two ways. First, they can give the impression that what is true of one colony is true for all colonies of that species. In other words, they can suggest that certain, stable attributes are characteristic of the species in general. Secondly, such observations can mask the presence of a dynamic process. Variation brought about by different conditions leads to an investigation of dynamics, the rules relating different states of colony behaviour to different conditions. Apparent uniformity leads to the assumption that there are no such dynamics, that only one state of the system is typical.

Johnston and Wilson (1985) expressed some surprise on finding significant variation among the caste distributions of four *Pheidole* colonies, and pointed out that inter-colony variation has received insufficient attention. To study the evolution even of caste distributions, it is clearly essential to know how they vary. Herbers (1983) goes further to point out, as I do here, that variation within colonies over time may be the consequence of interesting dynamics that have yet to be investigated.

For colony organization to evolve by natural selection, there must be variation in behaviour among colonies. The variation should be investigated. It is reasonable to consider, first, variation in those aspects of colony behaviour that affect reproductive success. As Oster and Wilson (1978) suggest, the amount of food a colony acquires, the extent to which it protects its nest and claims new territory, the numbers of new workers it produces to carry out its work—all can be expected to affect colony survivorship and reproductive success. The results discussed above show that these aspects of colony behaviour all depend on the dynamics that determine how foragers, patrollers, nurses, etc., react to each other and to changes in the colony's environment. Such dynamics are characteristic of species, but will vary among colonies of a species, as will environmental conditions. The evolution of colony behaviour involves the evolution of these dynamics. To understand the dynamics of colony organization, and how these dynamics evolve, students of social insect behaviour will have to adopt the experimental techniques of population biology. As well as looking at large numbers of individuals within colonies, we will have to look at large numbers of colonies, and how their behaviour changes with time.

2. THE EVOLUTION OF THE DYNAMICS OF SOCIAL INSECT BEHAVIOUR

Current theory of the evolution of social insect behaviour comes from two sources, kinship theory (Hamilton 1972), and the theory of the division of labour based on physical and temporal castes. Asymmetries in the relatedness of queens, males, and workers help to explain why a sterile worker class is perpetuated by natural selection. Kinship theory has also led to predictions about the sex ratios of colony reproductives (Trivers and Hare 1976). But kinship theory has not generated explanations of the complexities of the day-to-day behaviour of social insect colonies. On the other hand, optimization theory on the division of labour among worker subcastes *has* led to predictions about colony behaviour (Oster and Wilson 1978). However, the model is based on an assumption that colony behaviour is a function of caste distributions. This assumption is unrealistic.

Caste distributions are only a first step in predicting colony behaviour. In addition, the following factors must be considered:

1. For each task, baseline numbers of workers available to do it; caste distributions.

2. Switching rules: although individuals are committed to some baseline task, they may also switch tasks according to certain dynamic rules. These rules can be catalogued according to baseline task. For each group of workers, what tasks will they switch to, and under what conditions?

3. Intensity rules: at any moment, each of the colony's tasks has a certain group of workers committed to it. (This group may or may not contain workers that have switched over from some other task.) The intensity with which a given task is done, in terms of numbers of individuals carrying out the task and the amount of effort these individuals devote to it, depends on the dynamics of group-level interactions. That is, the intensity of effort devoted to a given task depends both on factors affecting how much of *that* task is done, and on factors affecting how much of *other* tasks are done. For each task, what causes the numbers engaged in it to increase and decrease? How do the numbers and effort devoted to a given task depend on the numbers and effort engaged in other tasks?

4. Developmental rules: switching and intensity rules change during the life history of the colony. How do the dynamics of (2) and (3) depend on colony age and colony size?

5. Variation among colonies in (1)–(4); in caste distributions, and in switching, intensity, and developmental rules.

The transition away from a research programme based on caste requires conceptual changes that are familiar in biology. The 'one gene–one protein' dogma of Beadle and Tatum was welcomed because it furnished a simple starting point for molecular biology. Jacob and Monod's discovery of the lac operon paved the way for new research which eventually showed that the dogma had to be abandoned. One cannot simply assign particular functions to particular genes; instead there are complex dynamics which regulate which genes are transcribed, when, and how much; and how the gene products function in the organism. In the same sense, though 'one worker–one task' provided a fruitful starting point for social insect research, it is now clear that more complex processes regulate what a worker does, when, how much, and what the colony-level consequences of a worker's behaviour will be.

We need new ways of understanding the organization of social insect colonies. In recent theoretical articles, Wilson and Holldobler (1988) and Seeley (1987) have emphasized two important aspects of colony organization. The first is that colonies function without hierarchically organized control systems. A version of this can be found in the Old Testament (Prov. 6:6) 'Look to the ant . . . who having no chief, overseer or ruler, provides her bread in the summer and gathers her food in the harvest'. The organization of social insect colonies is fundamentally different from that of hierarchically organized human groups, such as armies, corporations, orchestras, or governments. The behaviour of social insect workers is not directed by other insects in positions of authority.

A second point, discussed above, is that many of a colony's tasks are carried out in parallel, not in series. This is an efficient way to organize any system of interactive units, including social insect colonies, brains, and computers. There is an obvious analogy between brains and social insect colonies. Both are composed of relatively simple units, workers or neurons, that interact to do relatively complicated things: in colonies, maintain a nest, obtain food, and reproduce; in brains, think, learn, and remember. A long-standing debate in neurobiology centres around the question of whether particular functions are associated with particular neurons. Hebb (1949) and Lashley (1950) suggested that mechanisms for learning must be based on connections among neurons, and that particular neurons could play a variety of roles in the resulting interactions. At the time this seemed a radical suggestion, but it is now widely accepted (Dreyfus and Dreyfus 1988).

In neurobiology and artificial intelligence research, current interest in parallel processes extends to models that go a step further than the ideas outlined by Seeley and Wilson and Holldobler, to parallel *distributed* processes (Rummelhart and McClellan 1987). A system's functions are distributed when tasks are not assigned to specialized units, but are carried out by different units at different times. A parallel distributed processes (PDP) model of a social insect colony would attempt to describe the rules that structure the interactions of interchangeable workers, such as the rules governing changes of individual task and changes of colony activity. PDP models provide one way to take into account a third aspect of colony organization: colonies can behave predictably and adaptively without permanent individual specialization. Such models may be useful in formulating new theories of the evolution of social organizations not based on caste.

A PDP model of the dynamics of harvester ant behaviour (in preparation with B. Goodwin and L. Trainor) has several interesting properties. The predictions of the model can be formulated in terms of an energy landscape. The landscape sits in a space whose axes correspond to the numbers of ants engaged in various tasks. Any point or location in the landscape corresponds to a particular behavioural state of the colony, specified as the numbers of ants active in various tasks, and the numbers of ants inside the nest available to do each task. The height of any point in the landscape corresponds to an 'energy' value, a metaphor used to specify how likely it is for the colony to reach that state. An arbitrary convention, based on the idea of gravity, determines that colonies are more likely to be found in behavioural states of lower height or energy. In other words, colony behaviour tends to roll into the valleys of the landscape. Workers are assigned to different tasks, and can be either active or inactive participants in that task. Changes in the numbers of workers engaged in each task can come about in one of two ways: (1) individuals can switch tasks, and (2) the numbers engaged in various tasks can vary as individuals move back and forth between the active and inactive members of a particular task group. Such changes are determined by summing many simple interactions between pairs of ants, such as interactions between two active patrollers, or between an active patroller and an inactive forager, etc. The sign of the interaction determines its outcome. For example, a negative interaction between active workers of the same task group causes them to become inactive when they meet. If all permitted interactions are equal and negative, a system of simple negative feedback is set up. This can be adjusted to produce the stable distributions of individuals into task groups first suggested by early formulations of caste models.

The strength or magnitude of such interactions can also be varied. The relative magnitude of the interactions between different pairs of worker groups determines the bumpiness of the energy landscape. Smaller ratios produce a smoother landscape. A smoother landscape is more stable because the heights or energies of neighbouring regions of the landscape are similar.

This means that the probability of being in a particular behavioural state is close to the probability of being in a slightly different, neighbouring state. Suppose the interaction between foragers and patrollers is similar in magnitude to the interaction between nest maintenance workers and foragers. In this case, the dynamics of colony behaviour will be relatively stable. If interactions between the first two groups have little effect, while interactions between the second two produce large changes in the colony's behavioral state, colony behaviour will be much less stable. By varying the interaction strengths, it is possible to simulate the behavioural dynamics observed in harvester ant colonies. These results suggest that it may be useful to envisage the organization of behavioural flexibility in an ant colony as a parallel distributed process. Such models have provided interesting new questions about the flexibility that enables brains to learn and remember, and are now generating empirical hypotheses about the organization of social insect colonies.

New kinds of empirical research are needed to complement the formulation of new models of colony organization. Beginning a study of a social insect species, it is no longer sufficient to ask, how many tasks does this species do, and which individuals do each task? A study of flexibility may begin with these questions but must proceed towards some new ones. In a range of environmental conditions, how does this species' behaviour change? At least two kinds of change can be investigated. First, there are changes in task fidelity, the relation between individuals and tasks; that is, changes in which individuals do which tasks. Secondly, there are changes in the numbers of workers allocated to various tasks. Natural selection will act on variation within species in behavioural phenotype; thus variation in extent and type of flexibility, a crucial aspect of behavioural phenotype, must be measured. Understanding the relationship between variation in flexibility, and variation in survivorship, will open the way for ecological and evolutionary predictions. The ontogeny of these rules can be followed by testing how colonies of varying ages respond to alterations of their environments. In many species, mortality varies with colony age. This strongly suggests that ontogenic variation in a colony's reaction to its environment is a crucial life history characteristic.

There are a multitude of empirical questions to be addressed, analogous to those originally raised about caste distributions. We will need to know a great deal about the dynamics of colony behaviour: how species differ, how colonies of a species vary, how such dynamics change during a colony's life history, how they affect colony survivorship, and the extent of their heritability. The work reviewed above shows that considerable progress has been made towards addressing these questions. It is no longer possible to fit most empirical results into a theoretical framework that accounts for colony behaviour in terms of caste distributions. The theory of adaptive caste distributions inspired 20 years of productive work in social insect behaviour. The results call for new theoretical and empirical directions.

ACKNOWLEDGEMENTS

I thank D. Weild for helpful discussions; and P. Calabi, J. R. Gregg, R. L. Jeanne, E. A. Lloyd, R. Rosengren, E. O. Wilson, and the editors of this volume for their comments on the manuscript. The work was supported by a NATO postdoctoral fellowship from NSF.

REFERENCES

Bernstein, S. and Bernstein, R. A. (1969). Relationship between foraging efficiency and the size of the head and component brain and sensory structures in the red wood ant. *Brain Res.* 16, 85–104.

Calabi, P. (1987). Behavioural flexibility in hymenoptera: a reexamination of the concept of caste. In *Advances in myrmecology* (ed. R. J. Arnett Jr). E. J. Brill, New York.

——, Traniello, J. F. A., and Werner, M. H. (1983). Age polyethism: its occurrence in the ant *Pheidole hortnesis*, and some general considerations. *Psyche* 85, 395–412.

Carlin, N. F. (1981). Polymorphism and division of labor in the dacetine ant *Orectognathus versicolor*. *Psyche* 88, 231–44.

—— and Holldobler, B. (1983). Nestmate and kin recognition in interspecific mixed colonies of ants. *Science* 222, 1027–9.

Davidson, D. W. (1980). Some consequences of diffuse competition in a desert ant community. *Amer. Natur.* 116, 92–105.

Dreyfus, H. L. and Dreyfus, S. E. (1988). Making a mind versus modeling the brain: artificial intelligence back at a branchpoint. *Daedalus* 117, 15–44.

Fowler, H. G. and Roberts, R. B. (1980). Foraging behavior of the carpenter ant, *Camponotus pennsylvanicus* in New Jersey. *J. Kans. Ent. Soc.* 533(2), 295–304.

Gordon, D. M. (1984). The persistence of role in exterior workers of the harvester ant, *Pogonomrymex badius*. *Psyche* 91, 251–266.

—— (1986). The dynamics of the daily round of the harvester ant colony. *Anim. Behav.* 34, 1402–19.

—— (1987). Group-level dynamics in harvester ants: young colonies and the role of patrolling. *Anim. Behav.* 35, 833–43.

—— (in press). Dynamics of task switching in harvester ants. *Anim. Behav.* 38,

Hamilton, W. D. (1972). Altruism and related phenomena, mainly in social insects. *Ann. Rev. Ecol. Syst.* 3, 193–232.

Hebb, D. O. (1949). *The organization of behavior*. Wiley & Sons, New York.

Heinrich, B. (1979). *Bumblebee economics*. Harvard University Press, Cambridge, Massachusetts.

Herbers, J. M. (1977). Behavioral constancy in *Formica obscuripes* (Hymenoptera: Formicidae). *Ann. Ent. Soc. Am.* 70(4), 485–6.

—— (1979). Caste-based polyethism in a mound-building ant species. *Am. Midl. Nat.* 101(1), 69–75.

—— (1983). Social organization in *Leptothorax* ants: within- and between-species patterns. *Psyche* 90(4), 361–86.

Hölldobler, B. and Wilson, E. O. (1986). Ecology and behavior of the primitive cryptobiotic ant *Prionopelta amabilis*. *Ins. Soc.* **33**(1), 45–58.

Jaisson, P., Fresneau, D., and Lachaud, J. P. (1988). Individual traits of social behavior in ants. In *Interindividual behavioral variability in social insects* (ed. R. J. Jeanne). Westview Press, Boulder, Colorado.

Jeanne, R. L. (1986a). The organization of work in *Polybia occidentalis*: costs and benefits of specialization in a social wasp. *Behav. Ecol. Sociobiol.* **19**, 333–41.

—— (1986b). The evolution of the organization of work in social insects. *Monitore zool. ital.* **20**, 119–33.

Johnston, A. B. and Wilson, E. O. (1985). Correlates of variation in the major/minor ratio of the ant *Pheidole dentata*. *Ann. Ent. Soc. Ann.* **78**(1), 8–11.

Kolmes, S. A. (1985). A quantitative study of division of labour among worker honey bees. *Z. Teirpsychol.* **68**, 287–302.

Lashley, K. S. (1950). In search of the engram. *Symp. Soc. Exp. Biol.* **4**, 454.

Lenoir, A. (1979). Feeding behavior in young societies of the ant *Tapinoma erraticum*. L.: trophallaxis and polyethism. *Ins. Soc.* **26**(1), 19–37.

—— and Ataya, H. (1983). Polyethisme et repartition des niveaux d'activité chez la fourmi *Lasius niger* L. *Z. Tierpsychol.* **63**, 213–32.

Lindauer, M. (1961). *Communication among social bees*. Harvard University Press, Cambridge, Massachusetts.

McDonald, P. and Topoff, H. (1985). Social regulation of behavioral development in the ant *Novomessor albisetosus* (Mayr). *J. Comp. Psych.* **99**, 3–14.

Meudec, M. and Lenoir, A. (1982). Social responses to variation in food supply and nest suitability in ants (*Tapinoma erraticum*). *Anim. Behav.* **30**, 284–392.

Mirenda, J. T. and Vinson, S. B. (1981). Division of labor and specification of castes in the red imported fire ant *Solenopsis invicta* Buren. *Anim. Behav.* **29**, 410–20.

Moore, A. J., Breed, M. D. and Moor, M. J. (1987). The guard honey bee: ontogeny and behavioral variability of workers performing a specialized task. *Anim. Behav.* **35**, 1159–67.

Oster, G. and Wilson, E. O. (1978). *Caste and ecology in the social insects*. Princeton University Press, Princeton, New Jersey.

Plowright, R. C. and Plowright, C. M. S. (1988). Elitism in social insects: a positive feedback model. In *Interindividual behavioral variability in social insects* (ed. R. L. Jeanne). Westview Press, Boulder, Colorado.

Porter, S. D. and Jorgenesen, C. D. (1981). Foragers of the harvester ant, *Pogonomyrmex owheei*: a disposable caste? *Behav. Ecol. Sociobiol.* **9**(4) 247–56.

Rose, M. (1988). *Industrial behaviour*. Penguin, Harmondsworth, Essex.

Rosengren, R. (1977). Foraging strategy of wood ants (*Formica rufa* group) I. Age polyethism and topographic traditions. *Acta Zool. Fenn.* **149**, 1–30.

Rummelhart, D. E. and McClellan, J. L. (1987). *Parallel distributed processes*. MIT Press, Cambridge, Massachusetts.

Schmidt-Hempel, P. and Schmid-Hempel, R. (1987). Efficient nectar-collecting by honeybees. II. Response to factors determining nectar availability. *J. Anim. Ecol.* **56**, 219–27.

Seeley, T. D. (1986). Social foraging by honeybees: how colonies allocate foraging among patches of flowers. *Behav. Ecol. Sociobiol.* **19**, 343–54.

—— (1987). Foraging by honeybee colonies: a case study of decentralized control in insect societies. In *Chemistry and biology of social insects* (ed. J. Eder and H. Rembold), pp. 489–91. Verlag J. Peperny, Munich.

Smith, A. (1776). *The wealth of nations*. Penguin, London.

Sorensen, A. A., Busch, T. M., and Vinson, S. B. (1985). Behavioral flexibility of temporal subcastes in the fire ant, *Solenopsis invicta*, in response to food. *Psyche* **91**, 319–31.

Taylor, F. (1978). Foraging behaviour of ants: theoretical considerations. *J. Theor. Biol.* **71**(4), 541–66.

Traniello, J. F. A. (1978). Caste in a primitive ant: absence of age polyethism in *Amblyopone*. *Science* **202**, 770–2.

Trivers, R. L. and Hare, H. (1976). Haplodiploidy and the evolution of the social insects. *Science* **191**, 249–63.

Weiss, B. A. and Schneirla, T. C. (1967). Inter-situational transfer in the ant *Formica schaufussi* as tested in a two-phase single choice point maze. *Behaviour* **28**(3–4), 269–79.

Wilson, E. O. (1963). The social biology of ants. *Ann. Rev. Entomol.* **8**, 345–68.

—— (1968). The ergonomics of caste in the social insects. *Amer. Natur.* **102**, 41–66.

—— (1976). Behavioral discretization and the number of castes in an ant species. *Behav. Ecol. Sociobiol.* **1**(2), 141–54.

—— (1978). Division of labor based on physical castes in fire ants (Hymenoptera: Formicidae: *Solenopsis*). *J. Kans. Ent. Soc.* **51**, 615–38.

—— (1980). Caste and division of labor in leaf-cutter ants (Hymenoptera: Formicidae: *Atta*). I. The overall pattern in *A. sexdens*. *Behav. Ecol. Sociobiol.* **7**, 143–56.

—— (1984). The relation between caste ratios and division of labor in *Pheidole*. *Behav. Ecol. Sociobiol.* **16**(1), 89–98.

—— (1985). The sociogenesis of insect colonies. *Science* **228**, 1489–95.

—— and Holldobler, B. (1985). Caste-specific techniques of defense in the polymorphic ant *Pheidole embolopyx*. *Ins. Soc.* **32**(1), 3–22.

—— and —— (1986). Ecology and behavior of the neotropical cryptobiotic ant *Basiceros manni*. *Ins. Soc.* **33**(1), 70–84.

—— and —— (1988). Dense heterarchies and mass communication as the basis of organization in ant colonies. *Trends Ecol. Evol.* **3**(3), 65–8.

Winston, M. L. and Punnett, E. N. (1982). Factors determining temporal division of labor in honeybees. *Canad. J. Zool.* **60**, 2947–52.

—— and Fergusson, L. A. (1985). The effect of worker loss on temporal caste structure in colonies of the honeybee. *Canad. J. Zool.* **63**, 777–80.

Inclusive fitness in a nutshell

DAVID C. QUELLER

1. INTRODUCTION

Hamilton's theory of inclusive fitness (Hamilton 1964*a*,*b*, 1972) is one of the major underpinnings of the new field of sociobiology. Inclusive fitness theory describes how natural selection operates on a wide class of social behaviours: those in which the performer of the behaviour affects the fitness of one or more of its relatives. If the performer affects both its own fitness and that of a relative, changing them by $\Delta\omega_a$ and $\Delta\omega_b$ respectively, then the behaviour will be favoured by selection when a condition known as Hamilton's rule is satisifed:

$$\Delta\omega_a + r\Delta\omega_b > 0 \qquad (1)$$

Here, r is the relatedness between the two individuals. It has been defined in various ways, usually as genetic correlation or regression. It measures the relative likelihoods that the two individuals possess genes (above random levels) for the behaviour, so it serves to scale the relative importance of the two fitness effects. Hamilton's rule therefore quantifies the notion that a behaviour's fitness effects on relatives are important to the extent that the relatives are likely to possess genes for the behaviour.

Inclusive fitness theory has been widely applied in studies of animal behaviour, scoring successes in explaining both altruistic behaviour towards relatives and the mitigation of selfish behaviour among kin (e.g. Trivers and Hare 1976; Sherman 1977; Strassmann 1981; Reyer 1984). The principles involved are quite general, so it was to be expected that attempts would be made to extend the theory to other organisms. It has been recognized that inclusive fitness principles should also apply to plants (Hamilton 1964*b*; Nakamura 1980; Kress 1981), but in ways that are less obvious than in animals. Two preconditions must hold if kin selection is to operate in plants. First, individual plants must be able to affect the fitnesses of other plants. Surely they can do so in some circumstances, but their immobility and lack of rapid behaviour limits them in comparison with animals. Secondly, they must be able to direct their effects towards relatives, and plants are not known to have sophisticated kin recognition mechanisms (except for mating incompatibilities).

However, even in plants, there is at least one kind of interaction among individuals where these limitations do not apply: the process of raising offspring. Trivers (1974) was the first to explore systematically how inclusive fitness theory applies to parent–offspring relations. The salient conclusion from his work is that parents and offspring will often be in conflict with each other. For example, he noted that both a mother mammal and her offspring may be able to affect the amount of milk provided to the latter. An offspring could increase its own fitness by demanding extra milk, but it would do so at the cost of depriving its siblings. According to inclusive fitness theory, this trade-off should be seen quite differently by the mother and her offspring. To the mother, it is a trade-off between two offspring to whom she is equally related. But from the perspective of a selfish offspring, it is a trade-off between a close relative (itself, $r = 1$) and a more distant one (a sibling, $r \leqslant 1/2$). A conflict is therefore expected. Any given offspring should try to obtain more milk for itself than its mother should want to give it, and phenomena like weaning conflict are expected to evolve.

2. PARENT–OFFSPRING CONFLICT IN PLANTS

If plant embryos, growing on their mother, can affect the amount of resources received from the mother, then the principles outlined above ought to apply (Cook 1981; Westoby and Rice 1982; Queller 1983; Uma Shaankar *et al.* 1988). Elaborate recognition mechanisms are not required because attachment to the maternal plant serves as a reliable cue of the relationship. However, in the seed plants, the situation is more complicated because there are more kinds of participants involved. Recall that seed plants evolved from ancestors with alternating gametophyte (haploid) and sporophyte (diploid) generations. In the seed plants, the gametophyte generation is reduced and entirely dependent on its parent sporophyte. Therefore the gametophyte's offspring, the new generation of diploid sporophytes, also begin their development on the maternal sporophyte. The situation in gymnosperms is therefore rather like that in diploid viviparous animals, except that a multicellular haploid phase is intercalated between the diploid parent and offspring. The haploid gametophyte is an intermediary in more than a temporal sense. It is the resource acquisition and storage tissue in gymnosperms, sequestering nutrients that are later used by its embryos.

In the angiosperms, or flowering plants, the situation is still more complicated (Fig. 1). The same three generations are involved, but the gametophyte is even more reduced and has lost its acquisition and storage function. Though there are variations, it usually has only eight identical haploid nuclei. One is the egg nucleus, which is fertilized to create the single embryo. However, two other gametophytic nuclei participate in a second fertilization event, fusing with a sperm nucleus identical to one that fertilized

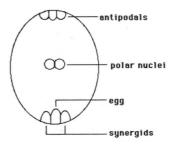

Fig. 1. Typical female gametophyte of an angiosperm. Eight identical haploid nuclei are included in an embryo sac, surrounded by diploid maternal tissue (not shown). One sperm nucleus fertilizes the egg to begin the embryo. A second sperm nucleus, identical to the first, fertilizes the two polar nuclei, creating a triploid nucleus that divides mitotically to make the endosperm. The remaining gametophytic nuclei generally develop no further and the role of nurse tissue is performed by the endosperm.

the egg. The product of this second fertilization is triploid and is identical to the embryo except that it has an extra complement of maternal genes. It divides mitotically to form the triploid endosperm. It is the endosperm that performs the acquisition and storage function in the flowering plants, although sometimes only during the early stages of development, after which it is devoured by the growing embryo. This is the Polygonum pattern of development typical of the majority of angiosperms (Maheshwari 1950). In a minority of species, there may be some differences in the number of maternal gene complements in the endosperm, and in some species with bisporic or tetrasporic development the maternal components may not be identical to each other or to the maternal complement of the embryo (Johri 1963). In this chapter, I will focus on the typical Polygonum pattern (see Bulmer (1986), for a discussion of the other types).

Thus there are four genetically distinct entities involved in the development of angiosperm seeds: the maternal sporphyte, the reduced haploid gametophyte, the triploid endosperm, and the embryo sporophyte. A number of authors recognized that inclusive fitness theory might prove useful in understanding this complexity (Charnov 1979; Krebs *et al.* 1980; Williams 1980; Willson and Burley 1983). Two questions immediately suggest themselves. First, how are the various genetic entities in the seed selected to behave if they can affect the amount of resources made available to their own embryo? Secondly, does this help us to understand the history of resource provisioning in the seed plants, particularly why the endosperm has replaced the gametophyte as the nurse tissue in the flowering plants?

A foundation for an inclusive fitness interpretation of the seed was developed independently by Westoby and Rice (1982) and Queller (1983). Its key feature is to extend Trivers' (1974) parent–offspring conflict theory by viewing each of the four genetic entities in the seed as a potential actor in the process by which the embryo of the seed is provisioned with nutrients. Endosperms (and gametophytes) that affect seed provisioning face the same trade-off that mothers and offspring do. By gaining extra resources for the embryo of their own seed, they decrease the resources available for embryos in other seeds of the same maternal plant. To understand how this trade-off is viewed from the standpoint of any actor (mother, father, embryo, gametophyte, or endosperm), we must know the relatedness of the actor to the individuals it affects: the embryo in its own seed ($r_{A,E1}$), and the embryos of other seeds on the maternal plant ($r_{A,E2}$). If the action of a provisioning gene in some actor causes a fitness benefit, b_{E1}, to the embryo of its own seed and a cost, c_{E2}, to other embryos on the same plant, a slight extension of Hamilton's rule (formula (1)) tells us that the gene should be favoured if $r_{A,E1}b_{E1} + r_{A,E2}c_{E2} > 0$. As before, each fitness effect is scaled by the relatedness of the actor to the individual affected. Rearrangement gives a form that is more easily compared for the various potential actors:

$$\frac{b_{E1}}{c_{E2}} > \frac{r_{A,E2}}{r_{A,E1}} \qquad (2)$$

Thus, any actor can be expected to attempt to get extra resources for the embryo of its own seed if the benefit–cost ratio exceeds some critical value determined by the relatedness. Table 1 gives the critical values derived by Westoby and Rice (1982) and Queller (1983) for each of the four potential actors in the seed (note that the earlier papers used inequality (formula (2)) in a different form, and used the reciprocals of the critical values shown in

Table 1
Relatedness ratios for possible participants in seed provisioning

Actor	$r_{A,E2}/r_{A,E1}$
Mother	1
Gametophyte	1/2
Endosperm	1/3
Embryo 1	1/4
Father	0

An actor should favour transfer of resources to its own embryo if the benefit divided by the cost to other embryos exceeds the relatedness ratio (see formula 2). $r_{A,E1}$ is the actor's relatedness to the embryo of its own seed and $r_{A,E2}$ is the relatedness of the actor to other embryos of the same maternal plant. Each embryo is assumed to have a different father.

Table 1; the choice is arbitrary). I have also included the father of embryo 1 (or the paternal gametophyte) as one of the potential participants. The values presented here assume that the embryos on the same maternal plant are outbred half-siblings, that is, that they have different fathers, unrelated to the mother. The rank order of these threshold values does not change if some of the embryos are full-siblings, so I will retain the half-sibling assumption throughout this chapter.

Table 1 shows that inclusive fitness theory makes some clear predictions about which genetic entities should fight hardest on behalf of their own seed. The mother has no particular 'own' seed: she is equally related to all her embryos and should favour transfer of resources from one to another only when the benefit exceeds the cost. (Note that this does not mean that she should provision them all equally; aborting an embryo is favoured when the benefits to other embryos are large enough). As was argued by Trivers (1974) for animals, an embryo's interests differ from its mother's. It should attempt to get more than its mother is selected to give. Specifically it should favour more resources for itself whenever the benefit–cost ratio exceeds $1/4$. The two other seed tissues, which function as intermediaries between the mother and embryo, are also intermediate in terms of how they are selected. Both should favour their own embryo more than the mother does, but less than the embryo itself does. Of the two intermediaries, the endosperm should favour its own embryo more than the gametophyte should. Finally, if the father of the embryo could affect provisioning (which may be unlikely in most circumstances), it would favour its embryo at any cost to other (half-sib) embryos on the maternal plant.

There are two potential sources of confusion that should be dispelled immediately. One is whether it is appropriate to view a non-reproductive tissue like the endosperm as an evolutionary individual upon which selection can act. One answer is that it can reproduce. The contribution of inclusive fitness theory was the recognition that individuals can reproduce in ways other than producing offspring. A worker ant can lack ovaries and still produce indirectly by helping to rear its siblings. Similarly, an endosperm can reproduce if it acts to enhance the fitness of an individual that shares its genes. A deeper answer to the question would be that it is not really relevant whether we consider the endosperm as an individual. We can ignore the question and simply ask how selection operates on genes that are expressed in the endosperm. Inclusive fitness theory provides an answer.

How to calculate the genetic relatedness coefficients may be another source of confusion. Malécot's (1948) coefficient of kinship, Wright's (1922) correlation coefficient of relationship, and Hamilton's (1972) regression coefficient of relatedness have all been used. As it happens, any of these three can be used in this application (Seger 1981; Michod 1982). All three have Malécot's kinship coefficient for a numerator, and though the denominators differ for the three coefficients, they cancel out when in the relatedness ratio of formula

(2). Whichever measure is used, the result is a ratio that tells us the relative degrees to which the two kinds of embryos possess the actor's genes byond random expectation.

However, there is a related, more serious problem that will be the focus of most of this chapter. Any arguments erected on the basis of the relatedness ratios in Table 1 depend on the basic inclusive fitness model being either correct or at least a reasonably good approximation. Inclusive fitness models are not always exact. That is, their results do not always agree precisely with the results of exact genetic models. Sometimes the disagreement is relatively trivial. For example, strong selection will distort pedigrees and alter the values of relatedness coefficients relative to what is expected under the assumption of no selection, but such effects are usually small (Toro *et al.* 1982). The development of exact genetic models for kin selection in seeds has revealed some more fundamental problems with the use of genetic correlation or regression coefficients (Law and Cannings 1984; Queller 1984; Bulmer 1986). Therefore, before we can use inclusive fitness theory as a basis for a theory of selection in seeds, we need to understand what these genetic models are telling us. Most of the remainder of this chapter is devoted to the task of revealing the causes and consequences of the departures from simple inclusive fitness behaviour in seeds. It turns out that there are two effects that need to be understood, and that these same two effects crop up in other uses of inclusive fitness theory. A better understanding of the fascinating yet idiosyncratic angiosperm seed can therefore also serve as the kernel from which a better understanding of inclusive fitness will grow.

3. EXPRESSION-DEPENDENT RELATEDNESS

One cause of deviations from simple inclusive fitness behaviour is non-additivity of the costs and benefits of a behaviour. This will be treated later in the chapter. For the time being, matters will be simplified by the assumption that all fitness costs and benefits are additive.

Inclusive fitness effects occur when a social behaviour is performed and not when the behaviour is left undone. It might seem unlikely that this statement, which borders on a truism, could reveal anything of real interest about inclusive fitness. But in fact it can. The key is to realize that the genetic correlation and regression coefficients that we have used contain no information about whether the behaviour of interest was performed. When our interest is in interactions among sibling embryos, we have implicitly assumed that it is sufficient to know the average pedigree relatedness among such embryos and that this will reflect the important genetic relations that pertain when the behaviour is actually expressed. This assumption is in fact often valid, but not always. When it fails, I will say that relatedness is expression-dependent.

One of the most useful features of the genetic coefficients is that they depend solely on pedigree connections. If these coefficients are valid, we do not need to know any details about how the social phenotype is expressed. This is no longer true when relatedness is expression-dependent. For example, we will see that relatedness sometimes comes to depend on the genetic dominance of the trait. We would still like to have at least a qualitative result that holds for all modes of expression. To this end, a major goal of the rest of this chapter will be to see if the qualitative rankings obtained from the strictly genetic coefficients are generally valid. Is it generally true that the rankings obtained from Table 1 are correct, even if the relatedness coefficients sometimes turn out to depends on the peculiarities of dominance? But first, I will review two examples of the general phenomenon of expression-dependent relatedness: inbreeding and parent-specific gene expression.

3.1   Inbreeding

The chief example in the literature arises because of inbreeding (Michod 1979; Seger 1981; Uyenoyama 1984; Grafen 1985). Under inbreeding, an individual's two alleles are not inherited independently. One result is that homozygotes have different genetic connections to their kin than do hetero-zygotes (Seger 1976). When homozygotes have different kinship patterns from heterozygotes, we might expect it to be advantageous for individuals to evolve social behaviours that are conditioned on their genotypes (e.g. if I am a homozygote, do A, if I am a heterozygote, do B, see Seger 1976). This kind of behaviour would require individuals to assess their likelihood of being homozygous, and this may not usually be possible. However, there is another manner in which the differences between homozygotes and heterozygotes can be manifested. A recessive gene will be expressed only in homozygotes, and its selective effect will therefore depend only on the kinship relations that pertain to homozygotes. A dominant gene will be expressed in both homozygotes and heterozygotes, so its selective effect will depend on both kinds of kinship pattern. As it happens, the genetic regression coefficient is correct under inbreeding only when dosage is additive so that heterozygotes as likely as homozygotes to express the behaviour (Seger 1981). But as a general rule, the expression-dependent relatedness under inbreeding make the purely genetic coefficients incorrect. This is because the genetic coefficients do not take phenotypes into account.

Michod and Hamilton (1980) developed a modified relatedness coefficient that makes Hamilton's rule work under inbreeding:

$$r = \frac{\text{Cov}(P,G')}{\text{Cov}(P,G)} \tag{3}$$

Like the simple genetic coefficients, it includes measures of the genes possessed by the actor, G, and its partner, G'. These measures can be interpreted either as breeding values of a quantitative trait or as frequencies of an allele at a single locus (taking values of 0, $\frac{1}{2}$, or 1). But in order to accommodate the expression-dependence of relatedness under inbreeding, eqn (3) also includes the phenotypic value of the potential actor, P, which we will assign as $P=1$ when it performs the behaviour and $P=0$ when it does not. This covariance formula is more transparent when it is transformed in the following manner (which is permissible if the mean values of G are the same in potential actors and their partners):

$$r=\frac{\sum\limits_{P=1}(G'-\bar{G})}{\sum\limits_{P=1}(G-\bar{G})} \qquad (4)$$

This is similar to a form derived by Grafen (1985). Now, the relatedness coefficient can be seen to consist of the ratio of two simple quantities. The denominator is the sum of actor's genetic values, taken as a deviation from the population mean. The point is that it includes the genetic values of actual actors only, not all potential actors. It therefore summarizes the genes for the behaviour possessed by those who actually perform the behaviour. Similarly, the numerator summarizes the genes for the behaviour possessed by those whose partners perform the behaviour. It is therefore easy to see that the coefficient scales the relative importance of fitness effects on actual actors compared to fitness effects on their partners. This coefficient is very general, subject only to the assumption we made at the outset: additivity of costs and benefits.

I do not raise this point to deal with the issue of inbreeding in seeds, although it would certainly be germane to that issue. Rather, it is most relevant because, even in random mating species, an inbreeding-like phenomenon occurs in the typical triploid endosperms. These endosperms inherit two identical alleles through the maternal gametophyte and it should not be surprising if this non-random association of alleles has effects similar to the non-random associations arising from inbreeding. But before moving on to endosperms, consider a second arena in which relatednesses are expression-dependent.

3.2   Parent-specific gene expression

An individual's relatedness to its kin can be thought of as deriving from two sources, its mother and its father. In calculating a purely genetic relatedness we take an average of these connections, because mother and father contribute equally to the offspring's genotype (for autosomal genes). But consider a social trait that is expressed only when the appropriate gene is inherited from the mother and not expressed when the same gene is inherited from the father. Then the only genetic connections that are relevant are those that trace through the mother. Similarly, only connections through the father should be considered for any trait whose expression was completely dependent on inheritance through the paternal line. These kinds of traits exist (e.g. Barton *et al.* 1984; Surani 1987), although they must not be too common or Mendel's laws would never have been accorded the status of laws. They seem to require genes to have a kind of memory of their past history. This could be accomplished if maternal genes were somehow labelled in a way that paternal genes are not, such as by DNA methylation (Reik *et al.* 1987; Sapienza *et al.* 1987; Swain *et al.* 1987). Such genes have been implicated in some arguments about kin selection in seeds (Haig and Westoby in press). But my immediate goal is not to argue that such genes are important in themselves. Rather, they are of interest because a formal model used to describe them can also be applied to normal endosperm genes. The double dose of maternal alleles in the endosperm provides an asymmetry that makes the importance of the maternal and paternal pathways depend on simple dosage relations rather than some special memory mechanism.

3.3 Path analysis

Figure 2 illustrates the genetic connections among members of a sibship (labelled as Embryo 1 and Embryo 2) with a path diagram. For those unfamiliar with path diagrams (see Li 1955), a number along an arrow connecting two variables is the correlation coefficient between those two variables. In this diagram, all the variables except the one labelled P represent genes or breeding values. Thus, the genetic correlation between

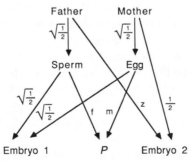

Fig. 2. Path diagram for calculating relatedness according to eqn (3). P is the phenotypic value of a trait affecting the fitness of embryo 1 and its half-sibling, embryo 2. The other variables (father, mother, sperm, egg, embryos) are breeding values. The arrows represent correlations between the variables they connect.

parent and gamete is $\sqrt{\frac{1}{2}}$ and that between parent and offspring is $\frac{1}{2}$. The genetic correlation between embryo 1's father and embryo 2 (z) depends on the probability that the father of embryo 1 is also the father of embryo 2. Path analysis provides a simple way to calculate correlations among two variables: (1) trace out each connecting path between two variables, with the only legitimate paths being first backward along arrows and then forward; (2) multiply the path coefficients in each path; and (3) sum the separate products obtained for each path. The genetic correlation between the two embryos is $\sqrt{\frac{1}{2}}\sqrt{\frac{1}{2}}(\frac{1}{2})+\sqrt{\frac{1}{2}}\sqrt{\frac{1}{2}}z$, which ranges from $\frac{1}{4}$ for half-sibs ($z=0$) to $\frac{1}{2}$ for full-sibs ($z=\frac{1}{2}$). These are the coefficients that apply as long as there is no expression dependence.

But we have seen that the more general relatedness coefficient, valid for expression-dependent cases, is given by eqn (3). If we want a general coefficient describing how selection operates when embryo 1 affects embryo 2, we need the covariances between the phenotype of embryo 1 with both its own genotype and with the genotype of embryo 2. Call these two quantities Cov($P,E1$) and Cov($P,E2$). Now let P in the path diagram represent the phenotype of embryo 1 for the social behaviour of interest, with m and f representing the contributions made to it by the genes from maternal and paternal gametes. As before, P takes on a value of 1 if the behaviour is performed, and 0 otherwise. The ratio of the two covariances is the same as the ratio of the corresponding correlation coefficients (because $r_{xy}=\text{Cov}(x,y)/\sigma_x\sigma_y$ and the two standard deviations are the same here). Therefore, using the rules of path analysis, and cancelling $\sqrt{\frac{1}{2}}$ in the numerator and denominator gives:

$$\frac{\text{Cov}(P,E2)}{\text{Cov}(P,E1)} = \frac{\frac{1}{2}m+zf}{m+f} \tag{5}$$

If m and f are equal, this turns out to be the same as the genetic coefficient, as expected when there is no expression dependence. If there is no paternal contribution to the phenotype ($f=0$), then only the path connections through the mother count, and relatedness is $\frac{1}{2}$ for both half- and full-sibs. If there is no maternal contribution ($m=0$), then only the connections through the father count, and relatedness ranges from zero for half-sibs ($z=0$) to $\frac{1}{2}$ for full-sibs ($z=\frac{1}{2}$).

Intermediate mixtures of maternal and paternal influence are also possible. This is exactly what we would expect to happen in the endosperm because it consists of an intermediate mixture of maternal and paternal genes. If we want to consider the effect of an endosperm phenotype (P) on its own ($E1$) and another embryo on the same maternal plant ($E2$), Fig. 2 and eqn (5) still apply (with the non-substantive substitution of male gametophyte for sperm and female gametophyte for egg). If each of the three alleles in the endosperm contributes equally and independently to the phenotype, then the two maternal genes together contribute twice as much as the paternal one, and $m = 2f$. (The actual values of the path coefficients are not required, but in the absence of environmental variance contributing to P, they are $m = 2/\sqrt{5}$ and $f = 1/\sqrt{5}$, obtainable from the fact that the sum of the squared path coefficients contributing to a variable must equal one). Substitution into eqn (5) yields a relatedness coefficient of $\frac{1}{3}$ for the half-sib case ($z = 0$), which agrees with the genetic coefficient of Table 1. As in the inbreeding example, the genetic coefficient is correct under additive dosage. This is because under additive dosage, each allele has the same effect on phenotype, regardless of its history. The histories can therefore be averaged over all alleles, and this is essentially what the genetic coefficients do.

But that is not the end of the story. Recall, that the presence of two identical maternal genes in the endosperm is somewhat analogous to inbreeding and in the inbreeding case, the genetic coefficient was correct only under additive gene action. The assumption that $m = 2f$ amounts to an assumption of additive gene action (each of the three alleles contributes equally to P). Here, as in the inbreeding case, non-additive dosage effects can lead to expression-dependent relatednesses. What would the appropriate relatedness be for endosperm phenotypes that are expressed by certain genotypes and not by others?

Consider selection on an allele, A, which when expressed in the endosperm, affects the fitness of its own embryo and other embryos on the same plant. For the purposes of clarity, restrict attention to the half-sib case ($z = 0$). Suppose first that the behaviour is expressed only when the endosperm has genotype **Aaa**. As the A must have been inherited through the father (any allele inherited through the mother would have two copies), the endosperm's behaviour is determined entirely by paternal genes. The relevant genetic connections are through the paternal line only ($m = 0$, $f = 1$), so the relatedness ratio is zero. Suppose now that the behaviour is expressed only in **AAa** endosperms. Now only maternal genes affect the behaviour ($m = 1$, $f = 0$) and the relatedness ratio $\frac{1}{2}$. These considerations show clearly that expression-dependence can occur in the endosperm, with results that can mimic the effect of embryo genes that are only expressed if they have been inherited through one particular parent. Of course, these two particular modes of expression are not particularly likely, but similar principles apply to other simple modes of expression, tabulated in Table 2.

If A is recessive and the endosperm behaviour is expressed only by **AAA** genotypes, then both maternal and paternal lines contribute equally ($m = f$) and the relatedness ratio is $1/4$. The fact that there are two maternal A's does not alter the conclusion that mother and father contribute equally to the phenotype because the two maternal A's are not inherited independently and do not have independent effects on the phenotype. For the endosperm behaviour to be expressed, it is necessary and sufficient for the mother to contribute an A and for the father to contribute an A, a completely symmetrical condition.

Table 2
Expression-dependent   relatedness   ratios   for endosperms

Endosperm phenotype is **A** when:			Relatedness
Aaa	**AAa**	**AAA**	ratio
X			0
X		X	0
		X	1/4
X	X		1/4
X	X	X	1/4
	X		1/2
	X	X	1/2

The relatedness ratio is calculated from eqn (5).

Similar symmetries apply to two other cases. If the endosperm behaviour is expressed only by **Aaa** and **AAa** genotypes, then expression is dependent on either inheritance of **A** from the mother or inheritance of **A** from the father, so each pathway is equally important. If the behaviour is expressed by **Aaa**, **AAa**, and **AAA** genotypes (dominant **A**), the same conclusion applies because expression is dependent on either maternal inheritance, paternal inheritance, or both. In both cases the relatedness ratio equals 1/4. Thus, both dominant and recessive endosperm genes are selected as if they were embryo genes, as has been shown in previous models (Queller 1984; Bulmer 1986). This is not surprising when one considers that under these modes of expression, the second maternal allele has no influence on the phenotype, so that selection should be exactly like selection on the embryo which differs only by its lack of the redundant second maternal allele.

The two remaining cases in Table 2 can be understood with similar principles. When expression is limited to **AAa** and **AAA** genotypes, termed '2-threshold' expression (Law and Cannings, 1984), inheritance of the **A** through the maternal gamete is necessary and sufficient for expression. Whether the third (paternal) allele is **A** or **a** has no influence on expression, so the relatedness is 1/2. This is the same ratio that pertains to the maternal gametophyte (Table 1) because the endosperm expresses the phenotype in exactly the same conditions as the gametophyte would: inheritance of **A** from the mother (Queller 1984). Finally, consider the rather unlikely mode of expression in which only **Aaa** and **AAA** endosperms express the phenotype. Here a paternal **A** is necessary and sufficient, so only connections through the paternal genotype are counted, and the relatedness ratio equals zero.

3.4   A general formulation

Queller (1984) derived a general formula for the relatedness ratio, which embraces the modes of expression discussed above and intermediate cases of partial expression. Here I will derive a general formula in a different way, which makes the underlying causes clearer. As eqn (5) is generally valid, all we need in addition are general expressions for the path coefficients m and f when P is considered to be the phenotype of the endosperm. That is, we need the correlations between endosperm phenotypes and: (1) the genotype of the maternal gamete; and (2) the genotype of the paternal gamete. Calculating a general expression for these correlations directly is rather complex, but it can be simplified by realizing that a correlation between any two variables is equal to the standardized covariance between the two variables, that is, the covariance calculated after standardizing each variable so that its mean is zero and its standard deviation is unity. The covariance of standardized variables X and Y is simply ΣXY.

Standardizing the genotypic values of maternal and paternal gametes is simple. One subtracts the mean and divides by the variance. If alleles **A** and **a** are at frequencies p and $1-p$, this means subtracting p and dividing by $p(1-p)$ (I am assuming random mating). Therefore if the gametic genotype is **A**, with an unstandardized value of 1, the standardized value is $G=(1-p)/p(1-p)=1/p$. If the gametic genotype is **a**, with unstandardized value 0, the standardized value is $G=(0-p)/p(1-p)=-1/(1-p)$. Now we need standardized values for endosperm phenotypes. Let us simply write them as H_0, H_1, H_2, and H_3 for endosperms with 0, 1, 2, and 3 **A** alleles respectively. We could have written each in more complicated form, explicitly showing the manipulations performed on the unstandardized values (H_3, for example, would be the frequency at which **AAA** genotypes perform the behaviour, minus the population frequency, divided by the variance). But this extra work turns out to be unnecessary because standardizing leaves the relative distances between the variables unchanged, and it is these relative distances that turn out to be important. We can now write expressions for the maternal and paternal path coefficients as

$$m=\Sigma HG_m = H_0 \frac{-p}{1-p}(1-p)^2 + H_1 \frac{-p}{1-p}p(1-p) + H_2 \frac{1}{p}p(1-p) + H_3 \frac{1}{p}p^2$$

$$f=\Sigma HG_f = H_0 \frac{-p}{1-p}(1-p)^2 + H_1 \frac{1}{p}p(1-p) + H_2 \frac{-p}{1-p}p(1-p) + H_3 \frac{1}{p}p2$$

Each consists of four terms pertaining to endosperms with 0, 1, 2, and 3 **A** alleles. Each term consists of the standardized phenotypic value of the endosperm, the standardized genetic value of the maternal or paternal gamete, and the frequency of the endosperm type. Note that the path coefficients are identical except for the genetic values in the terms representing **Aaa** and **AAa** endosperms (the H_1 and H_2 terms). Simplification of these equations gives:

$$m = (1-p)(H_2 - H_0) + p(H_3 - H_1) \tag{6}$$

$$f = (1-p)(H_1 - H_0) + p(H_3 - H_2) \tag{7}$$

Substitution of these values into eqn (5), along with $z = 0$ (for half-sibs), leads to an equation for the relatedness ratio that confirms the results in Table 2 and also works for other intermediate modes of expression:

$$\frac{\text{Cov}(P,E2)}{\text{Cov}(P,E1)} = \frac{\frac{1}{2}[(1-p)(H_2 - H_0) + p(H_3 - H_1)]}{(1-p)[(H_2 - H_0) + (H_1 - H_0)] + p[(H_3 - H_1) + (H_3 - H_2)]} \tag{8}$$

Equation (8) can be shown to encompass equation 12 in Queller (1984), where the probabilities of endosperms expressing the behaviour were assigned as 0, h_1, h_2, and 1 for genotypes **aaa**, **Aaa**, **AAa**, and **AAA**. However, the causes behind the strange endosperm behaviour are now more apparent. The relatedness ratio that actually applies to any mode of expression depends on two things. The first is that the endosperm's genetic connections to the embryos it affects differ through the maternal and paternal lines (except in the full sib case). This much is unremarkable; it also applies to the embryo. What is exceptional is that the relative importance of these two pathways can vary even without some special memory mechanism that tells a gene which parent it came from. The extra dose of maternal genes provides an asymmetry that permits simple dosage relations to determine the relative importance of the maternal and paternal connections. The formulas for the path coefficients (eqns (6) and (7)) provide a simple way to appreciate this. If an endosperm inherits a maternal **A** allele, it will gain two **A** alleles. Its effect will be to change endosperm expression from H_0 to H_2 when the paternal allele is **a** $(1-p$ of the time) or from H_1 to H_3 when the paternal allele is **A** (p of the time). Similarly, if an endosperm inherits a paternal **A** allele, it will gain one **A** allele. The effect will be to change endosperm expression from H_0 to H_1 when the maternal allele is **a** $(1-p$ of the time) or from H_2 to H_3 when the paternal allele is **A** (p of the time).

We can now return to the question that motivated this whole discussion. To what extent can the simple genetic coefficient of $\frac{1}{3}$ be used as reliable guide to endosperm behaviour? Clearly, it is not always correct. Because dosage affects the relative importance of the maternal and paternal relatedness, it is possible to find modes of gene expression that result in relatedness ratios that range from the paternal gametic one of zero to the maternal gametic ratio of 1/2. The value of $\frac{1}{3}$ is only recovered in the additive case, where each endosperm allele has the same average effect, regardless of its origin ($m = 2f$ or equal distances between each successive H-value).

Still, I think that the genetic regression coefficients have not been seriously misleading, at least given the assumption of additive costs and benefits that was made at the beginning of this section. The primary question of interest is where the endosperm stands in relation to the other entities that could affect resource allocation to seeds: the maternal plant, the maternal gametophyte, and the embryo itself. The strictly genetic coefficients predict an ordinal ranking of the degree to which these entities should favour investment in their own embryo at the expense of half-sib embryos. The embryo itself should be the most aggressive on its own behalf, followed by the endosperm, the gametophyte, and the maternal plant (Westoby and Rice 1982; Queller 1983). If the actual endosperm relatedness ratio usually falls between the gametophytic value of $1/2$ and the embryo value of $1/4$, then this ranking will generally be the correct one. Inspection of eqn (8) shows that this will be true as long as the probability of expressing the endosperm behaviour increases monotonically with the number of A alleles in the endosperm ($H_0 < H_1 < H_2 < H_3$). This is easiest to see by noting that this condition permits ($H_1 - H_0$) to range from zero to ($H_2 - H_0$) while ($H_3 - H_2$) can range from zero to ($H_3 - H_1$). Thus, as long as gene expression in the endosperm is not strictly dominant or strictly recessive, and as long as adding A alleles never decreases expression of the endosperm trait, then the ordinal ranking predicted by the genetic coefficients is correct.

This result can be understood a bit more intuitively as follows. When dosage is additive, each allele has the same effect on the phenotype, regardless of whether it came from the mother or the father. Therefore, in calculating relatedness, it is appropriate to average over the three alleles (that is, give double weight to the maternal pathway), and that is what the genetic coefficients do. When dosage is non-additive, paternally-derived alleles may have different average effects on phenotype than maternally-derived genes. The most extreme cases are strict dominance and recessiveness, when the second maternal allele has no additional effect. When this is true, it is appropriate to take an average over the maternal and paternal pathways, which gives the same value as averaging over alleles of the embryo. Intermediate dosage values will give relatedness values that are intermediate.

4. NON-ADDITIVE FITNESS COMPONENTS

The preceding discussion was based on the assumption that the fitness costs and benefits of the overconsumer behaviour are additive. Additivity means that the expected costs and benefits are independent of the number of costs and benefits one experiences. For example, the fitness cost to the brood of two embryos being overconsumers is assumed to be twice the fitness cost of

88   David C. Queller

having a single overconsumer. Additivity seems a reasonable assumption for many purposes, but it is not a completely general one. There are many circumstances in which the additivity assumption is inappropriate for behaviour among kin, including certain models of alarm calls (Maynard Smith 1965), warning coloration (Harvey *et al.* 1982; Guilford 1985), dispersal (Motro 1983), worker sterility (G. C. Williams and D. C. Williams 1957; Charlesworth 1978), and offspring competition for parental resources (Macnair and Parker 1979; Metcalf *et al.* 1979). These studies and others (Levitt 1975; Matessi and Jayakar 1976; Cavalli-Sforza and Feldman 1978; Templeton 1979; Uyenoyama and Feldman 1982; Queller 1985) have shown that inclusive fitness models using the standard relatedness coefficients are inexact with non-additive fitness components. We must therefore ask if non-additive fitness components are likely in seeds and, if so, does this change the qualitative conclusions arrived at through inclusive fitness considerations. In particular, does the predicted aggressiveness ranking still hold for the various seed tissues?

Before tackling the specific issues of seed competition, it is worth considering why non-additive fitness components disrupt inclusive fitness models. The reason is that non-additivity goes along with giving differential treatment to certain relatives based, not on pedigree relations alone, but on whether or not the affected relative also performs the behaviour. For example, in Charlesworth's (1978) model of the evolution of sterile workers in the social insects, he noted that sterile altruists exclude themselves from the class of potential beneficiaries of the altruism of others. The model is non-additive because sterile altruists have a direct fitness of zero no matter how many colony mates are altruistic. Altruists help only non-altruists, so the simple inclusive fitness model based on helping random colony members may fail.

This association between non-additive fitness effects and differential treatment of relatives is quite general (Queller 1985). Suppose performance of an altruistic behaviour subtracts c units from the altruist's fitness and adds b units to a sibling's fitness. Then if both perform the altruistic behaviour, their net fitness change may be written as $-c + b + d$, where d is the deviation from additivity of the two separate effects. If d is non-zero, it can be viewed as an additional increment or decrement to fitness caused by an altruist, but only when its partner is also altruistic. As altruistic siblings are a non-random set of siblings with respect to the altruism allele, inclusive fitness models based on average relatedness to siblings will not work. An inclusive fitness model can still be constructed for this case (Queller 1985) but it must take into account the additional genetic similarity that applies when both interactants are altruistic.

4.1 The model of Law and Cannings

The solution is not so simple in the case of interactions among developing seeds, because there are usually many seeds rather than just two, and there may therefore be many different non-additivities. Law and Cannings (1984) have developed a general model to describe such non-additive interactions. They define an overconsumer as an offspring that solicits extra resources and an underconsumer as one that does not. If we let the fraction of overconsumers in a brood be α, then the fitness of the overconsumers may be generally described as $\omega(\alpha)$ and that of the underconsumers $v(\alpha)$. Examples are shown as graphs in Figs 3–5. In principle, these functions could take any form, but Law and Cannings suggest that the following restrictions make the model most meaningful biologically:

$$\omega(\alpha) \geqslant v(\alpha) \tag{9}$$

$$\omega(1) = v(0) \tag{10}$$

$$\omega'(\alpha) < 0, \ v'(\alpha) < 0 \tag{11}$$

where the primes indicate derivatives of the functions. The first of these specifies that, within a brood, overconsumers have higher fitness than underconsumers. The second states that offspring fitness in a brood of all overconsumers is the same as in a brood of all underconsumers. For competition within broods for a fixed amount of resource, it can be viewed as specifying that there is no cost to being an overconsumer. If there is no such cost then both types of brood should divide resources in the same way. If there were a cost to overconsumption, it would decrease the net amount of resources available for the overconsumer brood, so we would expect $\omega(1) < v(0)$. The third set of conditions requires the derivatives of both

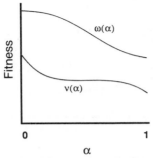

Fig. 3. Arbitrary fitness functions for overconsumer seeds, $\omega(\alpha)$, and underconsumer seeds, $v(\alpha)$. Both decline as a function of the fraction of the brood, α, that is composed of overconsumers

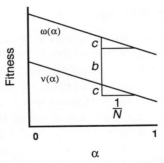

Fig. 4. Linear fitness functions for overconsumer and underconsumer seeds (eqns 12 and 13). The benefit to an individual of switching from underconsuming to overconsuming is *b*. The cost that it imposes on each sibling is *c*. Hamilton's rule is recovered for this case.

functions to be negative, meaning that adding overconsumers to a brood (increasing α) has a negative effect on the fitness of other brood members.

Using these fitness functions, and a single locus family model of selection, Law and Cannings (1984) derived certain ESS (evolutionarily stable strategy) conditions, some of which are shown in Table 3. The derivation need not concern us here, but I wish to point out two things. First, the conditions do not obviously match those from Hamilton's rule. Secondly, certain features of the inclusive fitness model discussed in the last section still hold. For example, the 2-threshold endosperm (one expressed by **aAA** and **AAA** genotypes) still behaves like the maternal gametophyte. Also, for both dominant and recessive genes, the endosperm behaves like the embryo. This should not be surprising since the reasons given in the last section still hold for any model. For example, if the endosperm expresses its overconsumption in exactly the same circumstances as the embryo would, as is true for both

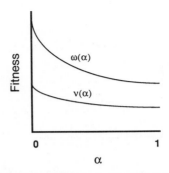

Fig. 5. Fitness functions for overconsumer and underconsumer seeds assuming that their fitnesses are proportional to the demands they make for resources (eqns (14) and (15)).

Table 3
Invasion criteria for overconsumer alleles of specified types. (Adapted from Law and Cannings 1984)

Where gene expressed	Type of expression	Invasion criterion	
Maternal tissue	Dominant or recessive	$\omega(1)$	$> v(0)$
Gametophyte	–	$\omega(\frac{1}{2})$	$> v(0)$
Endosperm	Dominant	$\frac{1}{2}\{\omega(\frac{1}{2}) + \omega(0)\}$	$> v(0)$
	2-threshold	$\omega(\frac{1}{2})$	$> v(0)$
	Recessive	$\omega(0) + v'(0)/4$	$> v(0)$
Embryo	Dominant	$\frac{1}{2}\{\omega(\frac{1}{2}) + \omega(0)\}$	$> v(0)$
	Recessive	$\omega(0) + v'(0)/4$	$> v(0)$

dominant and recessive genes, then selection should operate identically on the endosperm and embryo. Thus, the complicating effects of non-additive dosage remain in this model. If we want to examine the effect of non-additive fitnesses without these complicating effects, we should therefore examine the case of additive dosage, which was not considered by Law and Cannings (1984).

Table 4 lists the general boundary or ESS conditions for selection of additive overconsumer alleles in embryos, endosperms, gametophytes, and mothers. Two conditions are listed for each. The first is when an overconsumer allele can invade a population of underconsumers and the second is when an underconsumer can invade a population of overconsumers. I will illustrate the derivation of these conditions using the endosperm. First recall that additive dosage means that the probability of expressing the overconsumer trait is proportional to the fraction of alleles that are overconsumer alleles, so that **Aaa** endosperms are overconsumers $\frac{1}{3}$ of the time and **AAa** endosperms overconsumers $\frac{2}{3}$ of the time. Most of the population will consist of underconsumers in all-underconsumer broods, whose fitness is $v(0)$. If overconsumer alleles are to invade, this fitness must be exceeded by the fitness experienced by bearers of the overconsumer allele. When the allele is rare, nearly all bearers will be heterozygous **Aa** embryos resulting from **Aa** × **aa** matings. In half of these matings the mother will be heterozygous, and half of her offspring will be heterozygous with **AAa** endosperms, $\frac{2}{3}$ of which will be overconsumers. Therefore $\alpha = \frac{1}{3}$ for maternal broods, so the $\frac{2}{3}$ of the heterozygous offspring who have overconsumer endosperms get fitness $\omega(\frac{1}{3})$,

Table 4

General conditions for invasion of overconsumer and underconsumer alleles expressed in different tissues assuming brood members are half-siblings

	Overconsumer can invade when:	Underconsumer can invade when:
Father	$\omega(0) > (0)$	$v(1) > \omega(1)$
Embryo	$\frac{1}{2}[\frac{1}{2}\omega(\frac{1}{4}) + \frac{1}{2}v(\frac{1}{4}) + \frac{1}{2}v(0)] > v(0)$	$\frac{1}{2}[\frac{1}{2}\omega(\frac{3}{4}) + \frac{1}{2}\omega(1) + \frac{1}{2}v(1)] > \omega(1)$
Endosperm	$\frac{1}{2}[\frac{2}{3}\omega(\frac{1}{3}) + \frac{1}{3}v(\frac{1}{3}) + \frac{1}{3}\omega(0) + \frac{2}{3}v(0)] > v(0)$	$\frac{1}{2}[\frac{2}{3}\omega(\frac{1}{3}) + \frac{1}{3}v(\frac{1}{3}) + \frac{1}{3}\omega(1) + \frac{2}{3}v(1)] > \omega(1)$
Gametophyte	$\omega(\frac{1}{2}) > (0)$	$v(\frac{1}{2}) > \omega(1)$
Mother	$\omega(1) > v(0)$	$v(0) > \omega(1)$

while the $\frac{1}{3}$ that have underconsumer endosperms get fitness $v(\frac{1}{3})$. This accounts for the first pair of terms on the right-hand side of the formula. The second pair of terms is for the heterozygous offspring which have **Aaa** endosperms because their father was heterozygous. Only $\frac{1}{3}$ of these endosperms express the overconsumer trait and they are distributed one per brood (as we are assuming brood members have different fathers) so their fitness is approximately $\omega(0)$. The remaining $\frac{2}{3}$ are underconsumers who obtain fitness $v(0)$. The other boundary conditions in Table 4 are derived in a similar manner. The situation does not seem much improved; these conditions look no more like Hamilton's rule than those of Table 3.

4.2   Recovery of the additive case

However, if the model is correct, we ought to be able to recover Hamilton's rule at least for the special case of additive costs and benefits. This case can be represented by the linear fitness functions with equal slopes:

$$\omega(\alpha) = m\alpha + b_1 \tag{12}$$

$$v(\alpha) = m\alpha + b_2 \tag{13}$$

These functions are depicted in Fig. 4. A little geometry shows that this is equivalent to the additive case. Consider the effects of a single individual in a brood of N siblings switching from underconsumption to overconsumption. It will change the fraction of overconsumers from α to $\alpha + \frac{1}{N}$, so it changes its own fitness from $v(\alpha)$ to $\omega(\alpha + \frac{1}{N})$. Thus, the difference between these is labelled b (for benefit) in Fig. 4. This benefit to itself, in the limit of large N, becomes $\omega(\alpha) - v(\alpha)$, or in terms of our linear parameters, $b_1 - b_2$. The cost to its siblings arises from the fact that they now find themselves in a brood with $\alpha + \frac{1}{N}$ overconsumers instead of only α. Each overconsumer sibling suffers a fitness loss of $\omega(\alpha) - \omega(\alpha + \frac{1}{N})$ while each underconsumer suffers a loss of $v(\alpha) - v(\alpha + \frac{1}{N})$. In our equal-slope linear model these two costs are the same and are both labelled as c (for cost) on Fig. 4. Using the fact that the slope, m, equals the rise (c) over the run $(\frac{1}{N})$ the cost to each sibling can be written as M/N. The total cost to all $N-1$ siblings is therefore $(N-1)m/N$ which, in the limit of large sibships, approaches m. Thus, this linear model defines a constant benefit, $b_1 - b_2$, and a constant total cost, m, which applies to overconsumers and underconsumers equally.

If this is the additive case, substitution of the linear equations (12) and (13) into the conditions in Table 4 should yield the familiar inclusive fitness conditions from Hamilton's rule when dosage is additive. Table 5 summarizes the results, showing that Hamilton's rule is indeed recovered in each case. The same can be shown to be true for the conditions for non-additive dosage in endosperms in Table 2, with the proviso that the version of

Table 5

Conditions for invasion of overconsumer and underconsumer alleles with fitnesses of eqns (12) and (13)

	Overconsumer can invade when:	Underconsumer can invade when:
Father	$(b_1 - b_2) > 0$	$(b_1 - b_2) < 0$
Embryo	$\frac{1}{4}m + (b_1 - b_2) > 0$	$\frac{1}{4}m + (b_1 - b_2) < 0$
Endosperm	$\frac{1}{3}m + (b_1 - b_2) > 0$	$\frac{1}{3}m + (b_1 - b_2) < 0$
Gametophyte	$\frac{1}{2}m + (b_1 - b_2) > 0$	$\frac{1}{2}m + (b_1 - b_2) < 0$
Mother	$m + (b_1 - b_2) > 0$	$m + (b_1 - b_2) < 0$

Hamilton's rule recovered uses the phenotypic covariance measure of relatedness (eqn 3). Thus, the additive case is included in Law and Cannings' (1984) general model. But almost any departure from this equal-slopes linear model has two effects (see, for example, Fig. 3). The first is that the benefit, or the vertical distance between two lines, becomes variable and dependent on α. This means that the benefits are no longer additive. Secondly, the slopes of the two curves will differ, meaning the costs to overconsumers and underconsumers will be different. These are the reasons that Hamilton's rule fails in other models. The important questions here are (1) are such non-linearities likely to arise in cases of seed competition? (2) What are the consequences for how hard the various tissues will be selected to fight for their own embryo?

Previous models of offspring competition for parental resources have shown that the timing of offspring competition is critical in determining whether overconsuming behaviour may have different costs for overconsumer and underconsumer sibs. When overconsumers take resources that would otherwise have been used to rear more siblings in the future, then the costs will fall at random with respect to sibling genotype. Parker and Macnair (1978) constructed a model of parent–offspring conflict incorporating this assumption. Although its results are not expressed in terms of costs and benefits, they can in fact be converted into Hamilton's rule (Queller 1984). The same applies to models of gametophyte and endosperm selection. If the costs of overconsumption fall on later siblings, either directly (Queller 1984), or indirectly through reduced survival of the mother (Bulmer 1986), then Hamilton's rule is recovered.

However, if offspring compete simultaneously for a fixed amount of resource, then costs will not necessarily fall randomly with respect to genotype. Specifically, overconsumers gain extra resources, but purely at the expense of underconsumers. This is what causes Hamilton's rule to fail in models of simultaneous, or within-brood, offspring competition (Macnair and Parker 1979; Metcalf *et al*. 1979). As overconsumers exert a cost only on underconsumers, overconsumption is easier to evolve than is predicted by Hamilton's rule. While Law and Cannings (1984) constructed a general model that would incorporate endosperm and gametophyte behaviour, they did not substitute in any specific model of how offspring compete. I have shown that substitution of a linear model leads to the recovery of Hamilton's rule. Now, I will substitute a particular model of within-brood competition with differential costs.

4.3   The proportional solicitation model

Following Macnair and Parker (1979), assume that offspring (or endosperms or gametophytes) within a brood compete for a fixed amount of a maternal resource. Suppose that the mother allocates resource proportionally to the amount solicited and overconsumers solicit k times as much as underconsumers. Then in a brood of size N with αN overconsumers, the relative amounts of resources obtained will be $k/[k\alpha N+(1-\alpha)N]$ and $1/[k\alpha N+(1-\alpha)N]$, where the denominators are the total amount requested in a brood of N. If we assume for the moment that fitnesses are proportional to the amount of resource obtained, then relative fitnesses of overconsumers and underconsumers are:

$$\omega(\alpha)=k/[k\alpha+(1-\alpha)] \tag{14}$$

$$v(\alpha)=1/[k\alpha+(1-\alpha)] . \tag{15}$$

Note that for all $k>1$, these curves meet the conditions (eqns 9–11) considered to be biologically reasonable by Law and Cannings (1984). Equations (14) and (15) can be substituted into the boundary conditions given in Table 4 to find when each type can invade a population of the other type if overconsumers always get k times as much as resource as underconsumers. The results are listed in Table 6, with C set equal to zero (C represents a cost that will be added shortly).

The conditions do not look like Hamilton's rule and it is hardly worth trying to put them in a similar form when the assumptions underlying Hamilton's rule are so clearly violated (overconsumers benefit purely at the expense of underconsumers). Instead, I want to determine whether it is still valid to use the main qualitative conclusion obtained through use of Hamilton's rule, that ease of selection of overconsumer alleles follows the order: paternal > embryo > endosperm > gametophyte > maternal. The first thing to note is that the maternal conditions are neutral. This is as expected. Provided that there is no cost to overconsumption, mothers should be

Table 6

Conditions for invasion of overconsumer and underconsumer alleles with fitnesses of eqns (14) and (15)

	Overconsumer can invade when:	Underconsumer can invade when:
Father	$k > 1 + C$	$\frac{1}{k} > 1 - C$
Embryo	$(k^2 + 8k + 7)/(4k + 12) > 1 + C$	$(7k^2 + 8k + 1)/(12k^2 + 4k) > 1 - C$
Endosperm	$(k^2 + 10k + 7)/(6k + 12) > 1 + C$	$(7k^2 + 10k + 1)/(12k^2 + 6k) > 1 - C$
Gametophyte	$2k/(k + 1) > 1 + C$	$2/(k + 1) > 1 - C$
Mother	$1 > 1 + C$	$1 > 1 - C$

indifferent to whether their brood is all overconsumers or all underconsumers. But for the other participants, the inequalities in Table 6 are always satisfied for $k > 1$ (and $C = 0$). That is, overconsumption is always favoured by the father, embryo, endosperm, and gametophyte. There are two reasons for this. The first is that overconsumers gain purely at the expense of underconsumers. As the costs tend to fall on siblings who do not share the overconsumer gene, overconsumption is easier to evolve than Hamilton's rule would lead us to expect. The second reason concerns the relationship between the amount of food obtained and fitness. Strict proportionality has been assumed in this model. This means that if one seed were to give up one unit of fitness by not overconsuming, it would make one unit of fitness available to its sibling seeds. Thus, even if overconsumers affected siblings randomly with respect to genotype, so that Hamilton's rule should be valid, Hamilton's rule would still tell us that overconsumption would be favoured. Sacrificing one's own embryo's fitness for more distantly related relatives cannot be favoured unless the fitness sacrifice leads to a sufficiently larger fitness gain to the distant relatives.

If overconsumption is always favoured by embryos, endosperms, and gametophytes, then it would seem that they are not really in conflict with each other. But suppose that soliciting extra resources involves some extra cost, C, to the embryos of overconsumers. Subtracting C from the expression for overconsumer fitness (eqn (14)) means that the no-cost condition (inequality 10) no longer holds and, if C is large enough, it could even mean that overconsumers may have lower fitness than underconsumers (inequality 9 no longer holds). If we substitute the revised version of eqn (14) ($\omega(\alpha) = k/$

$[k\alpha + (1 - \alpha)] - C$), along with eqn (15), into the boundary conditions of Table 4, we obtain the results in Table 6. Now, for sufficiently large C, all of the invasion conditions can be met. It is now meaningful to ask which tissue is expected to be more aggressive on behalf of its own embryo. In other words, which is willing to undertake the largest cost in order to get additional resources?

Consider first the criteria for invasion of an underconsumer gene into an overconsumer population (Table 6). The smaller the left-hand side, the more easily the condition is satisfied (i.e. smaller C is required). Comparison of the left-hand side shows that ease of invasion of underconsumer genes follows the descending order: maternal, gametophytic, endosperm, embryo, father. This is exactly the order predicted by Hamilton's rule and it holds for all values of $k > 1$.

The situation is more complicated for invasion of overconsumer alleles (Table 6). The larger the left-hand side, the more easily overconsumer alleles invade (that is, they can invade in spite of larger C). For all $k > 1$, the left-hand side for the maternal condition is smaller than those for the other tissues so, as predicted, the mother is the least likely to favour overconsumption. Similarly, the left-hand side of the paternal condition is larger than all the others for $k > 1$. In addition, it is easy to show that the endosperm is less aggressive than the embryo, as predicted from Hamilton's rule.

The surprises come in the comparisons of these to tissues with the gametophyte. According to Hamilton's rule, the gametophyte should be the least aggressive of the three. But in this model, the gametophyte is less aggressive than the endosperm only if $2k/(k+1) < (k^2 + 10k + 7)/(6k + 12)$, a condition that simplifies to $(k-1)(k^2 - 7) > 0$. At $k = 1$, this condition is neutral, but as k increases from 1 the condition fails to be satisfied until we reach $k = \sqrt{7}$. Thus, for sufficiently small k, the gametophyte is predicted to be more aggressive. A similar condition holds for comparisons between the embryo and the gametophyte: the left-hand side of the gametophyte condition is larger than the left-hand side of the embryo condition for k in the range $1 < k < -1 + 2\sqrt{2}$. The gametophyte is therefore predicted to be more aggressive in this range. It is not hard to identify the reason why the gametophyte is sometimes most aggressive. Pedigree connections would still tend to make it less aggressive, but another force comes into play in this model. Recall that overconsumers gain purely at the expense of underconsumers. Although this is true under all four modes of control, the manner of expression makes a difference. As the gametophyte is haploid it always expresses the one allele it possesses. Gametophyte overconsumers therefore never impose a cost on siblings who share the overconsumer allele. In contrast, in our additive dosage model, some embryos and endosperms with the overconsumer allele act as underconsumers. Some of the cost of overconsumption therefore falls on individuals with the overconsumer allele, and this decreases the ease with which overconsumption is selected.

4.4 Summary

To summarize, the aggressiveness rank predicted by Hamilton's rule (father > embryo > endosperm > gametophyte > mother) is not always upheld in these models. When there is no cost to overconsumers ($C = 0$), then overconsumption is always favoured by the gametophyte, endosperm, embryo, and father. When there is a cost to the overconsumer, the ranking from Hamilton's rule is upheld for large k. But when k drops below $\sqrt{7}$, the gametophyte moves up one rank in aggressiveness, and when k drops below $-1 + 2\sqrt{2}$, the gametophyte is the most aggressive tissue. Bear in mind, however, that this applies only to the invasion criteria for overconsumer alleles. At the other end, the invasion criteria for underconsumer alleles (which is equivalent to fixation criteria for overconsumers) always follow the order predicted by Hamilton's rule. On the whole, then, the qualitative prediction from Hamilton's rule performs fairly well, but there are circumstances where it does not work. More work needs to be done in this area before any definitive answer can emerge. It would be worth exploring other models of simultaneous seed competition. At least two changes can be made in the model. First, resource allocation need not be proportional to the degree of solicitation. Secondly, fitness may not map linearly to the amount of resource obtained (although a linear approximation may be good for small differences in the amount of resource).

In the final analysis, the degree of confidence we can place in the simple inclusive fitness depends on which of the two models is more appropriate. The distinction has been characterized as being due to competition between broods and competition within broods, but it is really due to something more fundamental. If overconsumers affect their sibs at random with respect to the genes their sibs possess, then Hamilton's rule works well. However, when overconsumers do not affect siblings at random, but instead impose a cost on underconsumers alone, the aggressiveness ranking of the tissues may be different. It is therefore important to know which siblings bear the costs of overconsumption. This is an empirical question that will have to be answered by studies of particular plants. Modelling has suggested that the timing of competition is important. When overconsumption leads to decreased production of later siblings, its costs tend to be distributed at random, so Hamilton's rule works (Queller 1984; Bulmer 1986). When offspring compete simultaneously for resources, selection tends to lead to fiercer overconsumption, because only underconsumers are deprived of resources (Macnair and Parker 1979; Metcalf *et al.* 1979; Law and Cannings 1984).

But it should be recognized that, even with simultaneous competition, overconsumers may sometimes affect siblings at random. While there is no temporal dimension to ensure that a random subset of siblings is affected, spatial cues may perform the same role. Suppose, for example, there are several seeds in each fruit, each with an identifiable spatial position. Consider the fate of an allele coding the following behaviour: 'if I am in a seed at the peduncular end of the fruit, take extra resources that would otherwise go to other seeds in the fruit'. Here the action is dependent on a spatial condition that sorts the seeds into non-overlapping sets of potential actors (seeds in the peduncular position) and those who potentially bear the cost of overconsumption (the other seeds in the fruit). As the spatial position is determined environmentally (or at least randomly with respect to the overconsumer allele in question), the siblings who would bear the cost of the overconsumption are a random set of siblings. Even though this kind of allele affects contemporary siblings, its genetic effects are essentially the same as if it affected only later siblings, so Hamilton's rule should accurately describe how selection operates.

If we take both temporal and spatial dimensions into account, it seems likely that overconsumption will often affect siblings at random. In these circumstances, and also in some circumstances in which siblings are affected non-randomly, the aggresiveness of seed tissues should rank as: father > embryo > endosperm > gametophyte > mother. In some circumstances, however, this ranking may not hold perfectly. We can still make a weaker prediction that the father is selected to favour the most resources for its embryo, and the gametophyte, endosperm, and embryo are all selected to acquire more for their seed than the mother is selected to give. In this weaker prediction, the relative ranks of these three tissues are not resolved.

5. THE EVOLUTION OF THE ENDOSPERM

5.1   What needs to be explained?

Why is the endosperm of flowering plants constituted as it is, a genetic hybrid formed of unequal maternal and paternal contributions? Traditionally, hypotheses have centred on the possibilities for exceptional endosperm vigour owing to heterosis or polyploidy (Brink and Cooper 1940, 1947; Stebbins 1976). However, these ideas do not explain why these advantages are not met more simply through the use of maternal tissues (Westoby and Rice 1982; Queller 1983). Moreover, as they predate the application of kin selection to seeds, these hypotheses do not consider the possibility that increased vigour of the nurse tissue might actually be disadvantageous to the mother. I do not want to suggest that effects of heterosis and polyploidy are irrelevant to the endosperm's evolution, but in what follows I will focus on insights that might come from inclusive fitness considerations. Any explanation that fails to take the potential for conflicts of interest into account cannot hope to be complete.

Charnov (1979) was the first to propose hypotheses for the origin of the endosperm that gave a central role to conflicts of interest. His first hypothesis invokes kin conflict. Though he did not calculate relatedness ratios, he noted that double fertilization would increase the relatedness of the nurse tissue to its own embryo, relative to other embryos on the same plant (the real effect is due to a lowering of relatedness to the half-sib embryos). Charnov's second hypothesis involves intersexual conflict. Double fertilization may be a male strategy: by becoming part of the nurse tissue, a male can aid its own offspring at the expense of the offspring of other males. These two hypotheses have formed the basis for most subsequent work (Westoby and Rice 1982; Queller 1983; Willson and Burley 1983; Law and Cannings 1984; Bulmer 1986; Haig and Westoby, in press). As with all historical hypotheses, testing these ideas will be difficult. In this chapter I will focus on the preliminary task of exploring the logical strengths and weaknesses of the hypotheses, thereby determining which are the most serious candidates.

I will adopt a set of ground-rules for the analysis. I will assume that what needs to be explained is the transition from a haploid gametophytic nurse tissue to the Polygonum type endosperm, that is, the typical endosperm type that has been the focus of all analyses in this paper. This assumption is in accordance with the general view that the Polygonum endosperm type is primitive in the angiosperms (Johri 1963; Foster and Gifford 1974). This view is sometimes justified by the fact that the Polygonum type is the most common (Maheshwari 1950), a method of determining ancestral states that is no longer considered very useful (Estabrook 1977; Watrous and Wheeler 1981). But it can also be justified by outgroup comparison. Almost all gymnosperms have monosporic gametophytes so it is reasonable to assume that the first angiosperms did too (Law and Cannings 1984). However, a word of caution is in order here. Tetrasporic gametophytes (which are composed of descendants of all four meiotic daughter cells) have been reported in *Welwitschia* and *Gnetum* (Waterkeyn 1954; Martens 1963), which are thought to be among the closest living relatives of the seed plants (Doyle and Donoghue 1986). If the angiosperms evolved from a tetraporic ancestor, the exact scenarios expored below would not apply, but similar ones would.

Two pre-adaptations are required for the inclusive fitness hypotheses of endosperm origin to work. First, since they depend on the double fertilization resulting in changes in resource levels for the embryo, it must be true that provisioning is not yet complete by the time fertilization occurs. As angiosperms now provision their seeds after fertilization, this condition was clearly met before or during the transition to the flowering plants. The fact that gymnosperms lack an endosperm formed by double fertilization may well be attributable to this difference in the timing of provisioning (Westoby and Rice 1982; Queller 1983). The shift to later provisioning may have been selected to avoid wasting investment on seeds that either would not be fertilized or to allow maternal choice among seeds that were fertilized (Westoby and Rice 1982). Alternatively, Queller (1983) argued that the change might have been a basically non-adaptive consequence of a generally adaptive progenesis (bringing forward of mature traits, generally to speed up reproduction) thought to have occurred in the proto-angiosperms (Takhtajan 1969; note that he calls the process 'neoteny' rather than 'progenesis').

An additional preadaptation is likely to have been important. Many gymnosperm pollen grains deliver multiple identical sperm nuclei (Singh 1978; Wilson and Burley 1983). Therefore I will assume that multiple sperm nuclei were already available well before the evolution of the endosperm. We do not need to explain the delivery of multiple nuclei, only what happens after they are delivered. One other precondition is necessary. Our analyses assume that the endosperm helps its 'own' embryo, that is, an embryo that was fertilized by an identical sperm nucleus. This would be facilitated if relatively few pollen tubes, ideally only one, penetrate any given ovule (Queller 1983).

Within this framework, a good adaptive explanation of the origin of the endosperm should satisfy at least four criteria. First, it should account for both of the changes that have occurred in the transition from the haploid gametophytic nurse tissue: the fertilization by a male nucleus and the increased female contribution. I will assume that double fertilization was the first of the two steps because adding a second maternal nucleus prior to double fertilization has no effect on relatedness ratios. Secondly, it should explain whose adaptation it is. Was the endosperm's taking over of the nurse tissue role an adaptation of the mother, of the father, of the maternal gametophyte, of the embryo, or of the endosperm itself? This is an issue of gene expression and control: whose genes control the trait being selected? Thirdly, an adaptive hypothesis must say how the trait is adaptive. For the hypotheses discussed here, the adaptation will always be for taking either more or less nutrients for an embryo. Finally, another issue of control arises in cases of conflicts of interest. If the endosperm is an adaptation of one party in the seed to change the distribution of resources towards its optimum, at least one other party will be selected to oppose the change. We should like to know what determines why one party wins in spite of such opposition.

3.2   Intersexual conflict

Changes in the composition of the nurse tissue can have two kinds of effects. As I have emphasized throughout this chapter, one effect is to change the relatedness ratio for the nurse tissue, and thereby also change its preferred resource allocation. Of course, we cannot expect the newly constituted tissue immediately to change to its new preferred distribution. It does not 'know' its changed relationship to embryos. Only many generations of selection on genes expressed in the new tissue could allow it to act as if it knew its changed relatedness circumstances. Any explanation of the origin of the endosperm that relies on changed relatedness ratios must therefore account for this time-lag problem.

Changes in the composition of the nurse tissue may also have more immediate mechanical effects, and these can form the basis for selection if these effects happen to be adaptive. These are the kinds of changes assumed in the intersexual conflict hypotheses (Law and Cannings 1984; Bulmer 1986). Bulmer's genetic models have defined when some of these changes are favoured. First, he showed that, for a gene expressed in pollen, double fertilization to form a diploid nurse tissue (identical to the embryo) would be favoured if it resulted in more investment going to the embryo. This is what would be expected from the father's relatedness ratio of Table 1 and it provides formal support for Charnov's (1979) hypothesis that double fertilization is a male strategy that adaptively affects investment in the male's embryo. But why would adding a male nucleus to the nurse tissue lead immediately to increased nutrients being gathered for the embryo? Bulmer (1986) suggests that heterosis may be responsible, and notes that the heterosis theory of Brink and Cooper (1940, 1947) is compatible with the sexual conflict theory.

The weakest point in this scenario is what comes after double fertilization has evolved. If the male has succeeded in getting excess provisions for its embryo, against the interests of the female, why does the maternal plant allow double fertilization to continue? By preventing the second fertilization, she could restore the nurse tissue to its less aggressive haploid form. Instead, the response is hypothesized to have been the addition of a second dose of maternal genes. Bulmer (1986) shows that this is favoured (for genes expressed in the maternal gametophyte) when the effect of adding the extra maternal dose is to cause the nurse tissue to take less resources for its embryo. That is, the second dose of maternal genes is a maternal counter-adaptation. However, Bulmer also notes that we might normally expect that the immediate effect of an increase in ploidy will be to make the tissue more, rather than less, vigorous. This part of the scenario therefore seems weaker.

Haig and Westoby (in press) have suggested one way in which the increase in maternal ploidy could automatically lead to a less vigorous nurse tissue. First, they assume that once an embryo (or an embryo-like endosperm) is able to affect provisioning to itself, selection will lead to parent-specific gene expression. Specifically, if paternally-derived embryo genes could determine their paternal origin they would be selected to take more resources. In response, maternally-derived genes would be selected to take less. The end result would be an embryo (or endosperm) whose overconsumer alleles would be expressed primarily when derived from the father. In this circumstance, adding an extra dose of maternal genes would not be expected to increase expression of overconsumer genes. Of course, it would not necessarily be expected to decrease expression of these genes either, but it might if a second assumption of Haig and Westoby is met. They suppose that the total expression of endosperm genes is limited by something, perhaps competition for nucleotides, polymerases, or amino acids. Doubling the maternal contribution to create a triploid tissue would not, as noted above, increase the number of overconsumer alleles being expressed, but it would increase the number of other kinds of genes being expressed. The active overconsumer alleles would then constitute a smaller fraction of all active genes, so they would get relatively less of the limiting factor, and produce less of their gene product. The seed would therefore get less resources. This hypothesis provides a plausible adaptive explanation for the second dose of maternal genes, but its assumptions need to be tested.

3.3 Kin conflict

The alternative to the intersexual conflict hypothesis invokes kin conflict. It gives importance to the adaptive strategies of the new endosperm itself, not just the maternal and paternal plants. Charnov (1979) noted that adding a male nucleus to the haploid nurse tissue alters its relatedness ratio in a way that changes its optimal provisioning strategy. Specifically, the new endosperm (which at this point, prior to adding the extra maternal genes, is identical to the embryo) should favour more investment for its embryo. This hypothesis has been elaborated in a variety of ways (Westoby and Rice 1982; Queller 1983; Willson and Burley 1983). As noted above, the major obstacle to this kind of hypothesis is that the newly constituted endosperm cannot immediately 'know' what its optimal strategy is. Instead, it must be subject to a number of generations of selection before we can expect endosperm genes to behave as their relatedness ratio would predict. In other words, this explanation alone cannot explain the origin of the endosperm. But it might still explain why the endosperm takes over the gametophytic nurse role (Queller 1983). But to understand this, it is necessary to clear away what seems to be a common misconception about how double fertilization must have started.

Most discussions of the origin of the endosperm seem to assume, usually implicitly, that double fertilization and the endosperm nurse tissue arose simultaneously. After all, fertilization of the cell that would lead to the haploid nurse tissue will make that nurse tissue diploid. But I know of no reason to assume that, when double fertilization first arose, the second sperm nucleus fertilized the single cell that was the primordium for the entire nurse tissue. Extant gymnosperms have multicellular haploid gametophytes. Even if there was considerable reduction of the gametophyte in the line leading to angiosperms, it seems more likely that the second fertilization would be of one nucleus out of many. The remaining gametophytic cells would be unfertilized and would probably continue to develop into nurse tissue, perhaps along with the fertilized portion. Thus, double fertilization does not by itself eliminate the gametophyte. It seems likely that two nurse tissues, one fertilized and one not, would persist until one was eliminated by selection. Queller (1983) explained how kin selection would lead to the elimination of the gametopytic nurse tissue. Suppose that the original resource distribution reflected the gametophytic optimum. A new nurse tissue resulting from double fertilization would be selected to acquire more for its embryo (see Table 1). The gametophytic nurse tissue has a simple response: acquire less. With the fertilized tissue constantly being selected to take more and the unfertilized one continually responding by taking less, the end result would be the withering of the gametophytic nurse tissue. Eventually it would have reduced its role so much that it would not be acquiring and storing any resources.

This hypothesis has one particularly attractive feature. It explains why double fertilization is allowed to continue in spite of the fact that it goes against maternal interests. Once the fertilized nurse tissue begins taking too much resources from the maternal point of view, a maternal selective response is expected. It seems to me that the easiest response is the one described above. Instead of preventing the second fertilization, which might be difficult to accomplish (i.e. the required genetic variation may not be readily available), the response can be a simple decrease in the amount of resources acquired by the gametophyte. Selection is likely to adopt the most expedient response even if it is not the best long-term response. In this case, the gametophyte gradually gives up its resource acquisition function, leaving the endosperm in charge. At this point, preventing the second fertilization is no longer a viable response because then there would be no nurse tissue.

There is a variant of this hypothesis that is perhaps a bit stronger. Suppose that the double fertilization was originally nothing more than the fertilization of a second egg cell, so that the ovule has two identical twins and a haploid gametophyte (Willson and Burley 1983). The second twin is selected to help its sibling if there is a net gain in fitness, that is, if it gives up less fitness than its sibling gains. From this point, the scenario described in the preceding paragraph follows, with the second embryo gradually taking over the nurse function from the gametophyte. This variant of the hypothesis would help explain why the double fertilization occurred in the first place and why its product persisted for long enough to allow selection to give it its own strategy.

The weak point of the kin selection argument is accounting for the second dose of maternal genes. If the mother doubles her contribution to the endosperm and the new triploid endosperm persists for long enough to evolve its own strategy, it will favour taking less resources for its embryo than the diploid endosperm (compare endosperm and embryo in Table 1). This is to the advantage of the mother. But it is not clear why it persists for long enough to evolve its own strategy.

Westoby and Rice (1982) have developed a special version of this hypothesis. They note that it could be advantageous for the mother to select among her newly fertilized embryos, choosing those that demonstrate the best combinations of genes. Their hypothesis has the advantage of explaining why investment is deferred until after fertilization. But putting the embryo in a position to affect its own resource acquisition allows it to try to acquire too much from the mother's point of view. Westoby and Rice suggest that the triploid endosperm is a maternal compromise. It allows her to test particular combinations of maternal and paternal genes, but in a tissue that is less selfish than the embryo. For this hypothesis to work, it is necessary that the maternal advantage gained through the decreased selfishness of the endosperm exceeds the disadvantage due to the poorer quality of information provided by the endosperm (testing for good embryo gene combinations is best done on the embryos themselves). Moreover, creating the endosperm does not get rid of the selfish embryo. For the hypothesis to work, the mother must somehow be able to ignore the continuing demands of the embryo, paying attention only to the endosperm.

To summarize, inclusive fitness considerations do seem to provide plausible explanations for the origin of the endosperm. Both the intersexual conflict hypothesis and the kin conflict hypothesis can provide adaptive explanations for why the male contributes to the nurse tissue. I would give a slight edge to the kin conflict hypothesis because it accounts more clearly for failure of the mother or the gametophyte to evolve the counteradaptation of preventing the second fertilization. The explanations for the extra dose of maternal genes have more weaknesses, but not necessarily fatal ones.

5. CONCLUSIONS

Inclusive fitness theory provides a way to understand selection on seed provisioning. The picture of the seed that emerges is a complicated one, with several genetically distinct parties involved, each trying to enforce its own preference as to how much should be invested in the seed. I have not discussed the empirical evidence bearing on the application of inclusive fitness arguments to seeds. The approach seems to be supported by a number of facts, particularly from plant embryology (Cook 1981; Westoby and Rice 1982; Queller 1983; Willson and Burley 1983; Haig 1987, Haig and Westoby 1988; Uma Shaankar et al. 1988). In general, maternal conflict with offspring tissues (endosperm and embryo) seems fairly well supported by phenomena like maternal barrier tissues and offspring haustorial growths. The relative positions of the endosperm and embryo are much less clear. I should also point out that, although this may be the most elaborate application of inclusive fitness to plants, there have been others involving phenomena like pollen grouping, megaspore competition, polyembryony, seed dispersal, and also interactions among adult plants (Nakamura 1980; Kress 1981; Westoby 1981; Willson and Burley 1983; Ellner 1986; Haig 1986).

In this chapter I have concentrated on theoretical problems surrounding the evolution of seed provisioning, particularly the adequacy of simple inclusive fitness models. Two phenomena emerged as important. First, the asymmetrical contribution of the two parents to the triploid endosperm permits relatedness to be dependent on the mode of expression of the endosperm genes. Hamilton's rule is still correct if a more sophisticated measure of relatedness is used and, in any event, the qualitative predictions of even the simple versions of Hamilton's rule are generally unchanged. A more serious problem is encountered with non-additive fitness components, which are expected to arise in certain cases of simultaneous competition for resources. Here, certain qualitative predictions of Hamilton's rule may sometimes fail, but only sometimes, and only with regard to the relative aggressiveness of the gametophyte. I conclude that the predictions derived from simple genetic relatedness coefficients are usually reliable, at least as qualitative guides.

These predictions provide new insights into the evolution of the triploid endosperm. Several reasonable hypotheses have been devised, some invoking conflict between the parents, and others invoking the genetic interests of the endosperm itself. Although such hypotheses pass the test of theoretical plausibility, empirical testing will be more difficult. Most of these new ideas have been coming from workers with training in sociobiology, but we will need the additional expertise of plant palaeontologists, systematists, and embryologists to decide how well real world matches the theory.

ACKNOWLEDGEMENTS

I thank David Haig and Mark Westoby for permission to cite an unpublished manuscript. This paper was improved by suggestions from David Haig, Joan Strassmann, and an anonymous referee. This work was supported by the John Simon Guggenheim Foundation and by N.S.F. grants BSR86-05026 and BSR88-05915.

REFERENCES

Barton, S. C., Surani, M. A. H., and Norris, M. L. (1984). Role of paternal and maternal genomes in mouse development. *Nature* **311**, 374–6.
Boorman, S. A. and Levitt, P. R. (1980). *The genetics of altruism.* Academic Press, New York.
Brink, R. A. and Cooper, D. C. (1940). Double fertilization and the development of the seed in angiosperms. *Bot. Gaz.* **102**, 1–25.
—— and —— (1947). The endosperm in seed development. *Bot. Rev.* **13**, 423–541.
Bulmer, M. G. (1986). Genetic models of endosperm evolution in higher plants. In *Evolutionary processes and theory* (ed. S. Karlin and E. Noveo), pp. 743–63. Academic Press, Orlando.

Cavalli-Sforza L. L. and Feldman, M. W. (1978). Darwinian selection and altruism. *Theoret. Popul. Biol.* **14**, 263–81.

Charlesworth, B. (1978). Some models of the evolution of altruistic behaviour between siblings. *J. Theor. Biol.* **72**, 297–319.

Charnov, E. L. (1979). Simultaneous hermaphroditism and sexual selection. *Proc. Natl Acad. Sci. USA* **76**, 2480–4.

Cook, R. E. (1981). Plant parenthood. *Nat. Hist.* **90**, 30–5.

Doyle, J. A. and Donoghue, M. J. (1986). Seed plant phylogeny and the origin of the angiosperms: an experimental cladistic approach. *Bot. Rev.* **52**, 321–431.

Ellner, S. (1986). Germination dimorphisms and parent-offspring conflict in seed germination. *J. Theor. Biol.* **123**, 173–85.

Eastabrook, G. F. (1977). Does common equal primitive? *Syst. Bot.* **2**, 36–42.

Foster, A. S. and Gifford, E. M. (1974) *Comparative morphology of vascular plants.* Freeman, San Francisco.

Grafen, A. (1985). A geometric view of relatedness. In *Oxford surveys in evolutionary biology*, Vol. 2 (ed. R. Dawkins and M. Ridley), pp. 28–89. Oxford University Press.

Guilford, T. (1985). Is kin selection involved in the evolution of warning coloration? *Oikos* **45**, 31–6.

Haig, D. (1986). Conflict among megaspores. *J. Theor. Biol.* **123**, 471–80.

—— (1987). Kin conflict in seed plants. *Trends Ecol. Evol.* **2**, 337–40.

—— and Westoby, M. (1988). Inclusive fitness, seed resources, and maternal care. In *Plant reproductive ecology: patterns and strategies*, (ed. J. Lovett Doust and L. Lovett Doust), pp. 60–79. Oxford University Press.

—— and Westoby, M. (in press). Parent-specific gene expression and the triploid endosperm. *Amer. Natur.*

Hamilton, W. D. (1964a). The genetical evolution of social behaviour. I. *J. Theor. Biol.* **7**, 1–16.

—— (1964b). The genetical evolution of social behaviour. II. *J. Theor. Biol.* **7**, 17–52.

—— (1972). Altruism and related phenomena, mainly in social insects. *Ann. Rev. Ecol. Syst.* **3**, 193–232.

Harvey, P. H., Bull, J. J., Pemberton, M., and Paxton, R. J. (1982). The evolution of aposematic coloration in distasteful prey: a family model. *Amer. Natur.* **119**, 710–19.

Johri, B. M. (1963). Female gametophyte. In *Recent advances in the embryology of angiosperms* (ed. P. Maheshwari), pp. 69–103. International Society of Plant Morphology, Delhi.

Krebs, J. R. *et al.* (1980). Measuring fitness in social systems. In *Evolution of social behavior: hypotheses and empirical tests* (ed. H. Markl), pp. 205–18. Verlag Chemie, Weinheim.

Kress, W. J. (1981). Sibling competition and the evolution of the pollen unit, ovule number and pollen vector in agiosperms. *Syst. Bot.* **6**, 101–12.

Law, R. and Cannings, C. (1984). Genetic analysis of conflicts arising during development of seeds in the Angiospermophyta. *Proc. R. Soc. Lond. B* **221**, 53–70.

Levitt, P. R. (1975). General kin selection models for genetic evolution of sib altruism in diploid and haplodiploid species. *Proc. Natl Acad. Sci. USA* **72**, 4531–5.

Li, C. C. (1955). *Path analysis—a primer.* Boxwood Press, Pacific Grove, California.

Maheshwari, P. (1950). *An introduction to the embryology of angiosperms*. McGraw-Hill, New York.

Macnair, M. R. and Parker, G. A. (1979). Models of parent-offspring conflict. III. Intra-brood conflict. *Anim. Behav.* **27**, 1202–9.

Malécot, G. (1948). *Les mathématiques de l'heredité*. Masson, Paris.

Martens, P. (1963). Recherches sur *Welwitschia mirabilis*. III. L'ovule et le sac embryonnaire. *La Cellule* **63**, 309–29.

Matessi, C. and Jayakar, S. D. (1976). Conditions for the evolution of altruism under Darwinian selection. *Theoret. Popul. Biol.* **9**, 360–87.

Maynard Smith, J. (1982). The evolution of alarm calls. *Amer. Natur.* **94**, 59–63.

Metcalf, R. A., Stamps, J. A., and Krishnan, V. V. (1979). Parent-offspring conflict which is not limited by degree of kinship. *J. Theor. Biol.* **81**, 99–107.

Michod, R. E. (1979). Genetical aspects of kin selection: effects of inbreeding. *J. Theor. Biol.* **81**, 223–33.

—— (1982). The theory of kin selection. *Ann. Rev. Ecol. Syst.* **13**, 23–55.

—— and Hamilton, W. D. (1980). Coefficients of relatedness in sociobiology. *Nature* **288**, 694–7.

Motro, U. (1983). Optimal rates of dispersal. III. Parent-offspring conflict. *Theoret. Popul. Biol.* **23**, 156–68.

Nakamura, R. R. (1980). Plant kin selection. *Evol. Theory* **5**, 113–17.

Queller, D. D. (1983). Kin selection and conflict in seed maturation. *J. Theor. Biol.* **100**, 153–72.

—— (1984). Models of kin selection on seed provisioning. *Heredity* **53**, 151–65.

—— (1985). Kinship, reciprocity, and synergism in the evolution of social behaviour. *Nature* **318**, 366–7.

Reik, W., Collick, A., Norris, M. L., Barton, S. C., and Suranit, M. A. (1987). Genomic imprinting determines the methylation of parental alleles in transgenic mice. *Nature* **328**, 248–51.

Reyer, H.-U. (1984). Investment and relatedness: a cost/benefit analysis of breeding and helping in the pied kingfisher (*Ceryle rudis*). *Anim. Behav.* **32**, 1163–78.

Sapienza, C., Peterson, A. C., Rossant, J., and Balling, R. (1987). Degree of methylation of transgenes is dependent on gamete of origin. *Nature* **328**, 251–4.

Seger, J. (1976). Evolution of responses to relative homozygosity. *Nature* **262**, 578–80.

—— (1981). Kinship and covariance. *J. Theor. Biol.* **91**, 191–213.

Sherman, P. W. (1977). Nepotism and the evolution of alarm calls. *Science* **197**, 1246–53.

Singh, H. (1978). *Embryology of gymnosperms*. Gebrüder Borntraeger, Berlin.

Stebbins, G. L. (1976). Seeds, seedlings, and the origin of the angiosperms. In *Origin and early evolution of the angiosperms*, (ed. C. B. Beck), pp. 300–11. Columbia University Press, New York.

Strassmann, J. E. (1981). Kin selection and satellite nests in *Polistes exclamans*. In *Natural selection and social behavior* (ed. R. D. Alexander and D. W. Tinkle), pp. 45–58. Chiron Press, New York.

Surani, M. A. H. (1987). Evidences and consequences of differences between maternal and paternal genomes during embryogenesis in the mouse. In *Experimental approaches to mammalian embryonic development* (ed. J. Rossant and R. A. Pedersen), pp. 401–35. Cambridge University Press.

Swain, J. L., Stewart, T. A., and Leder, P. (1987). Parental legacy determines methylation and expression of an autosomal transgene: a molecular mechanism for parental imprinting. *Cell* **50**, 719–27.

Takhtajan, A. (1969). *Flowering plants: origin and dispersal*. Oliver & Boyd, London.

Templeton, A. R. (1979). A frequency dependent model of brood selection. *Amer. Natur.* **114**, 515–24.

Toro, M., Abugov, R., and Charlesworth, B. (1982). Exact versus heuristic models of kin selection. *J. Theor. Biol.* **97**, 699–713.

Trivers, R. L. (1974). Parent-offspring conflict. *Amer. Zool.* **14**, 249–64.

—— and Hare, H. (1976). Haplodiploidy and the evolution of the social insects. *Science* **191**, 249–63.

Uyenoyama, M. K. (1984). Inbreeding and the evolution of altruism under kin selection: effects on relatedness and group structure. *Evolution* **38**, 778–95.

—— and Feldman, M. (1982). Population genetic theory of kin selection. II. The multiplicative model. *Amer. Natur.* **120**, 614–27.

Uma Shaankar, R., Ganeshaiah, K. N., and Bawa, K. S. (1988). Parent-offspring conflict, sibling rivalry and brood size patterns in plants. *Ann. Rev. Ecol. Syst.* **19**, 177–205.

Waterkeyn, L. (1954). Études sur les Gnétales. I. Le strobile femelle, l'ovule et la graine. *La Cellule* **56**, 105–45.

Watrous, L. E. and Wheeler, Q. D. (1981). The out-group comparison method of character analysis. *Syst. Zool.* **30**, 1–11.

Westoby, M. (1981). How diversified seed germination behavior is selected. *Amer. Natur.* **118**, 882–5.

—— and Rice, B. (1982). Evolution of seed plants and inclusive fitness of plant tissues. *Evolution* **36**, 713–24.

Williams, G. C. (1980). Conclusion. In *Evolution of social behavior: hypotheses and empirical test*, (ed. H. Markl), pp. 205–18. Verlag Chemie, Weinheim.

—— and Williams, D. C. (1957). Natural selection of individually harmful social adaptations among sibs with special reference to social insects. *Evolution* **11**, 32–9.

Wilson, M. F. and Burley, N. (1983). *Mate choice in plants: mechanisms and consequences*. Princeton University Press, Princeton, New Jersey.

Wright, S. (1922). Coefficients of inbreeding and relationship. *Amer. Natur.* **56**, 330–8.

Levels of selection and sorting with special reference to the species level

ELISABETH S. VRBA

1. INTRODUCTION

Some of the most provocative evolutionary analyses since Darwin have shared an hierarchical approach, in that they proposed selection at levels other than the conventional one. These studies progressively broadened the traditional organismal focus to consider that not only organisms but also cells (Roux 1881; Weismann 1904; recently Buss 1987), demes (Wright 1932), genes (Williams 1966; Dawkins 1976), species (Eldredge and Gould 1972; Stanley 1979), and non-coding DNA sequences (Doolittle and Sapienza 1980) may be 'selfish'.

Why has the debate on levels of selection grown throughout all this time? If, as some have argued (reviews in Mayr 1978; Hull 1988), a theory about changes in gene frequencies and organismal adaptation by selection is by and large sufficient (barring peripheral exceptions), there is no need to continue this debate in the 1990s. All that needs resolution in such a case is the disagreement between a gene's and an organism's eye-view—which may largely correspond to equally valid (or equally flawed) perceptions of the visual illusion called the Necker Cube (Dawkins 1982). But instead of waning in importance, hierarchy, and levels of selection feature increasingly in evolutionary conferences and volumes (some are cited later in this chapter).

A number of reasons for this occur to me. First, we all know that life is structured hierarchically. The most exciting biological questions include how living things are able to build and to maintain a hierarchy of organizational levels. How did the entitites at different levels—replicating complexes, cells, eukaryotes, metazoans, species, among others—originate in the first place; and why did evolution so rarely dissemble them again into less inclusive forms? Interactions between selection at the old level and selection at the new emergent one *must* have played a part in the origin of each level. And the 'ghosts' of such ancient antagonistic and synergic interactions are still there today in living systems, as Buss (1987) argued with respect to the variety of metazoan life cycles. The need to understand such origins on its own constitutes sufficient reason to incorporate levels of selection in our theory.

Secondly, there is the related question of how the levels, such as those from DNA sequences to species in the genealogical hierarchy, have maintained

themselves and proliferated inside each other. Evolutionary theory has yet to
organize major aspects of the natural history of molecules, cells, development, sexual reproduction, and long-term lineages in a coherent way. Many
feel that certain evident relations between levels in terms of structure,
dynamics, and relative abundance remain poorly explained. Some such
evidence has been known in broad outline, and has bothered evolutionists,
for a long time. For instance, Maynard Smith (1976a) expressed the
bewilderment of many since Darwin (see also Bell 1982) at the prevalence of
sex: 'One is left with the feeling that some essential feature of the situation is
being overlooked'. There is a current groundswell of opinion (e.g. Arnold *et
al.*, in press) that a properly formulated expansion of the levels-of-selection
approach might illuminate such old enigmas.

Additional questions have been raised by the recent flood of new molecular and species-level data. Some who see these patterns in their laboratories in
effect say to the biological community: 'we need significant theoretical
expansion to explain these patterns'. Others are not so sure. Nevertheless, a
growing number from all sub-fields seem to agree that, as important causes
may indeed flow from processes at and between levels beyond those
traditionally addressed, we had better at least investigate the possibility.

1.1 What this paper is about

The traditional theoretical framework needs to be enlarged and refined to
study multilevel processes. Concepts that served well at the gene-organism
interface (and where, if used carelessly, rarely led to gross error), like
heritability, selection and adaptation, usually transfer with difficulty to other
levels, or cannot do so at all. For an internally consistent theory of multiple
levels such concepts need to be delimited appropriately and precisely in the
context of additional concepts (that we never had to, or cared to, imagine in
dealing with genes and organisms).

This chapter is a modest contribution towards this expanded conceptual
structure. I review distinctions and arguments of others, propose additional
ones that I regard as useful, and illustrate them with emphasis on the species
level. I see notable progress in the units-of-selection debate. Conceptual
distinctions that were fringe elements some 20, 10, and even five years ago,
have since been widely recognized as important. I will also discuss continuing
disagreements of three kinds. Semantic disagreements are the least important
kind. Distinctions between different concepts are crucial. Any real distinction
should be acknowledged by *some* name, if not a particular name. Other
differences may be of the Necker Cube kind in that alternative viewpoints
can each be reasonably advocated. For instance, Maynard Smith (1987)
argued that one may agree about the basic biology of a given situation, and
about all the results expected, yet variously represent it as a case of gene,
organismal, kin or group selection. Lastly, there are ontological disagree-

ments that do not disappear upon a different choice of terms or by a Necker Cube flip of viewpoint; and these are the most fundamental. But purely semantic or 'Necker Cube differences' in one context can result in ontological mistakes in a new context. Maynard Smith's (1987, p. 120) statement reflects this: 'Our choice of models, and to some extent our choice of words to describe them, is important because it affects how we think about the world ... [These choices decide] what phenomena we regard as readily explicable, and which need further investigation'.

2. A PRELIMINARY LOOK AT BASIC CONCEPTS

Several important concepts need brief clarification at the outset. As a general statement of **selection**, that applies at all levels of the genealogical hierarchy (but does not include 'meme selection', Dawkins 1982, or 'cultural selection', reviewed in Maynard Smith 1986), I suggest: **Selection is the interaction, between heritable, varying, emergent characters of individuals, and the common environmental elements experienced by the variant individuals, that causes differences in birth and/or death rates among them**. This statement is close to that in Vrba (1984), except that the notion of common environmental elements affecting the selected individuals has been added. I agree with Sober (1984) that this needs to be articulated.

Some comments on the terms used is required. Any entity that has a discrete beginning ('birth'), an end ('death'), and spatial localization, can be termed an **individual** (Ghiselin 1974; Hull 1980). Thus, genomic elements, cells, organisms, demes, species, and other living entities have in common that they are 'reproducing individuals'. I ask the reader's indulgence if I occasionally use terms such as 'reproduction', 'birth', and 'death' to apply at levels other than that of organisms: when one needs to refer collectively to similar processes at several levels, it is simpler to say, for example, 'reproduction at several levels' than 'gene replication, cellular division, organismal reproduction, the origin of a new deme from an ancestral deme, and speciation'. 'Reproduction' used in this sense means one individual dividing into two or more of similar characteristics, although possibly differing in some respects. 'Birth' and 'death' mean respectively the first and last appearance of an individual.

The term 'emergent' has been regarded with some suspicion among biologists. I hope to convince the reader that I have a pragmatic concept in mind: an **emergent character** is a structural or dynamic character that is level-specific in that it does not exist at any lower level. For instance, I have the emergent character of five digits on my hand, which is not a property of my cells or genes. An emergent character is in principle describable and measurable. Such an operational concept of emergence is the one referred to by systems scientists, for instance by Caswell et al. (1972, p. 39) in discussing mathematical modelling of dynamic systems: 'It is an unfortunate fact that emergent properties have sometimes been given an almost mystical character in the literature. They *are* real phenomena, which do arise to confront anyone studying a complex system'.

Animal and plant breeders routinely distinguish phenotypic selection from evolutionary response to selection (Falconer 1981). This distinction, emphasized by Haldane (1954), is still used by population geneticists:

> Natural selection acts on phenotypes, regardless of their genetic basis, and produces immediate phenotypic effects within a generation that can be measured without recourse to principles of heredity or evolution. In contrast, evolutionary response to selection, the genetic change that occurs from one generation to the next, does depend on genetic variation. (Lande and Arnold 1983, p. 1210)

My concept is restricted to Lande and Arnold's subset of selection that does depend on genetic variation.

Units of selection: I take the view that two sorts of entities function in each selection process: replicators (Dawkins 1978, 1982) and interactors (Hull 1980; Dawkins 1982 calls interactors 'vehicles', and Mayr 1988 'targets of selection'). In some cases, such as 'selfish DNA' selection (Doolittle and Sapienza 1980), the two functions are attached to individuals at the same level. In each case of higher-level selection there are two kinds of units of selection at two different levels. For instance, in species selection, species are 'interactor units of selection' by virtue of differences in species characters that interact with the environment to bias speciation and extinction rates; and the genes that are the heritable basis of the species characters are the 'replicator units of selection'. In the replicator-interactor distinction I follow Hull (1980, 1988, pp. 408–9), while my concept of selection (p. 113) differs slightly from his:

replicator—an entity that passes on its structure largely intact in successive replications;
interactor—an entity that interacts as a cohesive whole with its environment in such a way that this interaction causes replication to be differential;
selection—a process in which the differential extinction and proliferation of interactors cause the differential perpetuation of the relevant replicators.

Not all replicators influence phenotypes. By **'genes'** I mean only those genomic elements that have specific effects on phenotypes. When referring to other parts of the genome I will do so explicitly. There are two further distinctions. (1) 'Gene' refers to a molecule that bears hereditary information, and not to the information itself. That is, there may be many generations of genes inside a single organism in analogy with, for instance, generations of organisms in an asexual clone, without any change in the information content of the replicators in either case. (2) The term gene does not require reference to selection. In both respects I differ from Williams (1966, 1985). He uses gene to mean information and not the molecule itself;

and views a gene in evolutionary contexts as 'any hereditary information for which there is a favourable or unfavourable selection bias equal to several or many times its rate of endogenous change' (Williams 1966, p. 25). I agree with Williams on the need to separate the structure of the genomic material from the information it represents. For one thing, the former may change while the latter persists.

I use the term **evolution** to encompass all processes of change in heritable characters, and of changes in frequencies of such characters, at whatever level of the genealogical hierarchy they may occur. This view of evolution leaves open the question of causes. This again differs from Williams (1985, p. 21) who proposed a redefinition of 'evolution' in terms of selection: 'I suggest that the term evolution be used only in this sense, of a process determined by chance events plus the persistent biases of natural selection'. I fear that the exclusive 'hitching to the selection horse' of such time-honoured terms, that have so far been neutral with respect to particular causal process, would have a restrictive influence on the range of models that 'evolutionists' would be prepared to consider. In general, I will argue that we retain concepts that are not all defined in terms of the traditionally most popular ones, such as selection and adaptation.

3. 'LIFE IS ROUGH'

Biologists customarily single out a small segment of evolution for attempts at explanation. A molecular geneticist might explain the prevalence across particular taxa of a genomic sequence as selection of selfish DNA. A population geneticist could investigate in the same organismal group how inbreeding structure of populations affects a simple allelic system, and from this extrapolate to causes of speciation patterns. A palaeontologist may point to topographical heterogeneity and geographical distribution during the history of the same taxa as predominant causes of their differential speciation rates. Yet we all know that life does not evolve at one level at a time, keeping others conveniently constant. In reality, the long-term patterns at all three levels in our hypothetical taxonomic group may represent synergism, or antagonism with compromise, between simultaneous selection (and other causes) at all the three levels. **The processes at different levels all churn over together with causal forces propagating both upwards and downwards between levels, as lineages evolve nested inside each other**.

Thus, the cryptic title of this section has a double meaning: (1) the passage of life's lineages through time is not smooth and simple, but buffeted by interacting forces at and between several levels of complexity; and (2) this in turn makes life 'rough' for evolutionists in search of realistic explanations. There are several consequences that are of importance to the levels-of-selection debate.

First, most common traits that we see at any time level are probably favoured (or, where not favoured currently, at least tolerated) at all levels— products of long-term synergism between selective and other influences at different levels (as Buss 1987, personal communication pointed out). Thus, most often when a scientist concludes that a selective explanation of his narrow data set works very well, he is probably quite right up to a point. Selection at the given level has indeed been a factor, and he may be forgiven for assuming that his subject matter has overriding importance for under- standing evolution. This effect may account partly for why hierarchical approaches to evolution have been slow to be appreciated. It seems that we need to stumble on the rare ongoing selective conflicts (or dissonances arising from other factors) between levels to acknowledge the need for multilevel theory. For instance, many evolutionists do not consider group selection in cases that can be argued in terms of conventional selection, on grounds of parsimony, unless opposition between group and organismal selection is apparent. But as Wade (1977) and others pointed out, such explanation is off the mark if in fact both group and lower level selection are acting synergic- ally. The mistake would not be of the Necker Cube kind, but ontological.

Secondly, we may opt for a single-level selective explanation in a given case on grounds of evidence for the current absence of selection at an adjacent level. For instance, in the kinds of organisms on which most of evolutionary theory is based, the 'bottleneck life cycles' (each of which restarts with a single cell in each generation) prevent mutation and change in somatic cell lineages from reaching the next generation. Such metazoan ontogenies appear to be selected at the organismal level, with cell selection absent (apart from rare gametic selection and cancer) due to germ line sequestration. Yet I think that Buss (1987) is right: current ontogenies contain 'frozen' testimony of past selective interactions between cellular and multicellular levels. Evolutionary explanations need to address those ancient histories.

Thirdly, the view of selection regimes at different levels churning over together has implications for the significance of replicators in evolutionary theory. The same replicators must often simultaneously influence several interactors at different levels (e.g. Lewontin 1970; Templeton 1979; Yokoyama and Templeton 1980; Brookfield 1986). On the one hand this underlines the uniqueness and enormous importance of replicators: **only at the level of the replicator do such selection regimes have their evolutionary effects integrated**. Replicators, by their copying fidelity (Williams 1966; Dawkins 1978), provide the most reliable informational record (Williams 1985). On the other hand, **the replicator record, because it is the aggregated outcome of several different processes, cannot possibly provide adequate evolutionary explanation on its own.** (This has already been argued with respect to gene–phenotype interactions [e.g. Sober and Lewontin 1982]. Selection at additional levels renders the problem more cogent.) The heredit- ary emergent characters at the different higher levels, and the environmental contexts with which they interact, must be analysed as well in order to decipher evolutionary history.

In similar vein, the phenotypic frequencies and fitnesses, in cases where organismal and group (and/or species) selection act simultaneously, are aggregate responses to multilevel selection as well as additional factors (Sober and Lewontin 1982; Sober 1984). Thus, the criteria of group selection suggested by Arnold and Fristrup (1982), Wimsatt (1981), and Heisler and Damuth (1987), namely, that group membership or group attributes must account for some of the covariance between fitness and phenotypic traits, are insufficient to serve as the **diagnosis** of group selection.

4. THE CONCEPT OF SORTING

The notion of 'sorting' allows more freedom, with respect to causal processes implied, than does the selection concept. Some of us (Vrba and Eldredge 1984; Vrba and Gould 1986) have found this concept useful as a context within which the role of selection can be delineated from other processes.

Living entities at several levels share the characteristic of 'reproduction'. That is, they can give rise to 'offspring' that resemble them in general organization. The reproductive processes at different levels differ in their tendencies to error. Non-coding DNA sequences, mitotic cells, genes, and asexual organisms have the highest reproductive fidelity—for many generations they tend to reproduce identical descendants; that is, they replicate (Dawkins 1978). Sexual organisms routinely produce varied offspring; and 'giving birth' among species (speciation) relies crucially on the introduction of evolutionary novelty. Thus, the contemporary individuals (*sensu* Ghiselin 1974; Hull 1980) at each of these levels vary in characters. Reproduction and death among variant individuals result in some kind of association between birth- and death-rates and characters—a process often referred to simply as 'differential birth and death'.

Differential birth (or replication, cell division, or speciation) and death (or gene termination, cell death, or terminal extinction) are ubiquitous in living things, and feature in every evolutionary discussion. Therefore it is desirable that we have a short term for this collective phenomenon, and that the term be free of process implications beyond the birth and death processes themselves. The term 'sorting', in its vernacular meaning of grouping things with particular attributes in various combinations, has been suggested (Vrba and Gould 1986; Vrba and Eldredge 1984; Vrba 1984). Sorting may result in one or another correlation of birth- and/or death-rates with variant characters of the sorted individuals. It may also result in an uncorrelated pattern. (In the vernacular sense such a pattern may result, for instance, if one blindly removes pieces of differently coloured chalk from a bag and sorts them into piles of 10, with the piles varying randomly in colour frequency.)

118   Elisabeth S. Vrba

Thus, sorting is a simple description of differential birth and death; it contains in itself no statement about causes. **Darwin provided a theory for only one cause of sorting—natural selection acting upon heritable characters of organisms.** However, a number of other processes produce sorting as well. There are two distinctions of importance here. **A.** There are two kinds, or categories, of processes: the casual processes that produce sorting, and the sorting processes themselves. Selection belongs to the first category. **B.** There are several different kinds of causal processes that produce sorting. Selection is only one among these. Others will be discussed below (including random and non-heritable sorting, sorting by selection at a higher level ['hitch-hiker-sorting'], sorting by selection at a lower level [effect sorting], and sorting inside a higher entity by virtue of structure at the higher level). In the interests of clear communication about evolutionary causes and outcomes (Sober 1984), I urge that we need the descriptive concept of sorting—a concept that encompasses all manifestations of a ubiquitous process, differential birth and death, that applies to entities at different levels of organization, and that leaves open the question of causes.

5. SORTING IN RELATION TO FITNESS

There are different usages of the terms fitness and selection in the literature. As a result the characterization of fitness in relation to sorting and selection is not easy. As usually understood by evolutionists, fitness is a probabilistic quantity. Population geneticists commonly refer to alleles or genotypes as differing in fitnesses (see (2) below), meaning statistically expected results of natural selection. Where alleles are neutral (no selection acting), they have fitnesses but no fitness differences (although their actual reproductive histories may differ). One may also refer to fitness differences among heritable phenotypes resulting from selection, or to fitness differences among non-heritable phenotypes resulting from 'selection' (e.g. Lande and Arnold 1983; see quotation on p. 114). I summarize the population genetic applications as follows, using my concepts of selection (which is restricted to heritable characters) and sorting (which is a simple description of differential birth and death). Fitnesses include the expected outcomes of heritable sorting by selection or random drift, and the expected outcomes of non-heritable sorting; and can apply to alleles, genotypes, and phenotypes. In addition, other different versions of the term fitness have been and are still used. Dawkins (1982, pp. 181–7) gives a useful summary of five different usages of the term fitness:

(1) To Spencer, Wallace, and Darwin the fittest individuals meant, roughly, those with the highest capacity to survive or reproduce by virtue of features such as the keenest eyes or strongest legs.

(2) Population geneticists most often apply the word fitness to genotypes and alleles. The fitness W of a genotype **Aa**, for instance, is defined as $1-s$, where s is the coefficient of selection against the genotype. W is the average of the relevant probabilities of reproductive success attaching to the organisms that have the genotype.

(3) 'Classical fitness', as used by ethologists and ecologists, is a property of an individual organism, often expressed as the product of survival and fecundity.

(4) Hamilton's (1964*a,b*) inclusive fitness is a property of an organism's actions or effects. Inclusive fitness is calculated from an individual's own reproductive success plus his effects on the reproductive success of his relatives, each one weighed by the appropriate coefficient of relatedness.

(5) 'Personal fitness' of an organism in the sense of Orlove (1975) 'may be briefly characterized as "the same as his fitness (3) but don't forget that this must include the extra offspring he gets as a result of help from his relatives" ' (Dawkins 1982, p. 187).

Sorting shares with these differing notions of fitness its relational nature: there is no sense in talking about sorting unless one is comparing the birth and/or death rates attached to variant individuals. Fitness (1) is strongly tied to natural selection of organisms. Fitness (2) describes well the expected outcome of sorting among 'gene individuals' (*sensu* Ghiselin 1974; Hull 1980) as a result of selection acting on them and on the organismal bodies they produced. (But bear in mind the range of usages by population geneticists outlined in the first paragraph of this section.) Fitness (3) provides the closest measure of sorting among organisms. The complex relations between the fitness (3) of an organism and the most commonly used fitness (2) of a genotype is well expressed by Dawkins (1982, p. 183):

The measured fitness (2) of the brown-eyed genotype will contribute to the fitness (3) of an individual who happens to have brown eyes, but so will the fitness (2) of his genotype at all other loci. Thus the fitness (2) of a genotype at a locus can be regarded as an average of the fitnesses (3) of all individuals possessing that genotype. And the fitness (3) of an individual can be regarded as influenced by the fitness (2) of his genotype, averaged over all his loci.

Fitness (4) and (5) each in a slightly different way focuses on the expected outcome of what may be termed 'inclusive sorting' as a result of selection of genes and phenotypes.

Thus, there is no unanimity about what fitness means. The most commonly used version, fitness (2), is defined in terms of selection and is tied to the levels of genes and genotypes. Other population genetic usages refer to fitnesses of heritable or non-heritable phenotypes. Taken together, the above versions of fitness imply a narrow range of causes with a strong emphasis on selection, and are restricted with respect to the levels addressed. In addition, fitness cannot be identified with actual reproductive history, namely sorting (Brandon 1978). As Sober (1984, p. 43) puts it: 'The census information of who [reproduces and] dies is evidence used for *estimating* fitness; actual [reproductive and] survival rates do not *define* fitness values'. It seems, therefore, that there is a separate role for a concept of sorting—a description that encompasses the actual differential birth and death processes among entities at whatever levels they may be situated, and that is agnostic on the question of causes.

When I refer to fitness, I will mean an expected outcome of differential representation as a result of sorting. I allow that selection acting directly on the entities in question is not the only possible cause of fitness differences of similarities among them. I generalize the notion by allowing that entities other than alleles, genotypes and organisms have fitnesses as well. For instance, group fitness refers to the expectation that groups with certain characteristics will produce new groups more often, or survive longer, than will groups with other characteristics (e.g. Maynard Smith 1987); and the terms species and lineage fitness have similarly been applied to expectations of differential speciation and/or extinction rates. This expansion simply lets fitness (3) above for organisms apply analogously at other levels.

6. A BASIC DISTINCTION: LOWER ENTITIES CAN BE SORTED WITHIN, OR BY SELECTION AMONG, HIGHER ENTITIES

Let us say that variants at a level are sorted non-randomly, and that selection at this level is not responsible. Instead, variation among higher entities dictates sorting at the lower level. There is a fundamental distinction, that is useful in considering levels of selection, between ways in which this can occur:

A,B. A property of some of the higher entities causes lower variants inside them to multiply disproportionately; although there is no sorting among the higher entities. That is, some higher entities grow 'fat' and others stay 'thin', while they retain the same fitness.

A. The 'fattening' of a higher entity occurs without change in frequency (sorting) among variants inside its 'body'. Still, as I will show below, there can be systematic sorting at the lower level when viewed across fat and thin higher entities.

B. The 'fattening' of a higher entity occurs with systematic sorting among variants inside its 'body'.

C. Selection at the higher level causes sorting at the lower level, without any sorting inside higher entities.

Although **C** is well known at several levels, the set including cases **A** and **B** (sometimes referred to below as set **A** + **B**) is unfamiliar. I will illustrate the three cases with reference to the simple diagrams in Fig. 1 (**A, B, C**). But before I do this, I want to point out that I do not assert that **A** and **B** are prevalent at several levels of the genealogical hierarchy, nor that either of them ever occurs at any level. Nor do I imply that they have not been discussed before, at one or another level and under some or other name. (In fact, I will argue below that some existing models do fit rather well.) I merely posit the theoretical distinction, which can at least in principle apply at different interfaces between levels, for exploration. Having generalized it, one can consider particular real-life case histories at different levels, or particular models in the literature, and ask: is this a case of **A**, **B**, or **C**, or of a combination **A–C**, or **B–C**?

6.1 Modes of sorting 'below' by dictates from 'above': a hypothetical illustration

Figure 1 represents the three kinds of evolution in three separate pairs of sister-lineages, **A, B**, and **C**. The lower entities (#'s and *'s) in each case can be thought of, for example, as organisms of two different phenotypes and the higher entities (the circles, ○'s) as groups. In each case the organisms undergo sorting between times 1 and 2, and we assume that the causes are neither random, nor selection among organisms.

In case **A**, a property of groups (ii) causes both lower variants inside them to experience the *same* increases in birth-rates, relative to what happens in groups (i). There is no sorting among groups—the only difference is that some are growing 'fatter' than others. There is no sorting (with change in relative frequency) among #'s and *'s within each group either. But growing 'fat' or staying 'thin' is characteristically associated with different relative frequencies of lower variants. Thus, when we look at the lower level as a whole, there is sorting going on.

In case **B**, every condition is the same as in **A** with one exception: The property of groups (ii), that causes both internal variants to multiply faster, also favours #- over *-phenotypes. Thus, this time there is sorting within each 'fat' group, although there is still no sorting among groups.

In case **C**, group properties result in group selection. There is no sorting within groups among the #'s and *'s. But group selection promotes # phenotypes overall.

Although ancestral frequencies of #'s and *'s, and their packaging in groups, are identical in **A, B**, and **C**, and although at time 2 after sorting the overall figures for #'s and *'s are also identical, the processes by which this occurred differ fundamentally. In **A** and **B** the #'s and *'s were sorted overall by processes that occurred *within the ancestral group (ii)*. There was *no sorting among groups*. In contrast, in case **C** there is *no sorting within groups*.

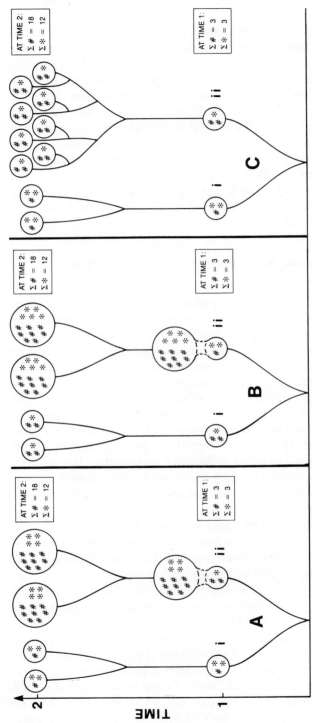

Fig. 1. Different ways in which lower entities (#'s and *'s) can be sorted by downward causation, or dictation, from higher entities (the circles, ○'s) in which they reside. For example, take the lower entities (#'s and *'s) to be organisms of heritably different phenotypes and the higher entities (○'s) to be groups. Cases **A**, **B**, and **C** represent three different lineages.

A and B: two versions of 'Mustapha Mond sorting'. Groups of kind **A**(ii) and **B**(ii) have emergent characters that cause organisms inside them to multiply disproportionately. In **A** this occurs without change in relative frequencies (sorting) of #- and *-phenotypes within group (ii). Still, there is sorting among phenotypes between times 1 and 2 when viewed across lineages (i) and (ii). In **B**, multiplication is accompanied by phenotypic sorting (# has higher fitness than *) in group (ii). In neither case **A** or **B** is there any sorting among groups.

C: group selection. Groups vary in emergent characters that result in group selection. There is no sorting within groups among #'s and *'s.

Sorting amoung groups, favouring ##∗ over #∗∗ groups, accounts for the outcome among #'s and ∗'s.

The same representations in Fig. 1A, B could also apply under different assumptions. The sorting of #- and ∗-phenotypes may be independent of which particular group it occurs in, or autonomous with respect to it. For instance, natural selection of phenotypes may differ between B(i) and B(ii) simply because groups (i) and (ii) happen to be located in differing environments. That is, here group structure has nothing to do with what happens.

In A and B the #'s and ∗'s were sorted overall by processes that occured *within the ancestral group (ii)*. Referring to 'sorting among focal entities that occurs by changes in their birth- and/or death-rates within the bodies of higher entities, as in Fig. 1A + B, and is not accounted for by selection at their own level but is instead dictated by the higher-level emergent properties', as I intend to do in further discussion below, is rather cumbersome. What shorter term could represent such processes at several levels of the genealogical hierarchy? The analogy of citizens within states may fit: to the extent that a national ruler or law dictates that members of the population with certain characteristics may have more children than others, or must die at different ages, sorting among humans depends on whether they live in that nation or in another more liberal one. In Aldous Huxley's (1946) *Brave new world*, the differential representation of human genotypes was 'downward dictated' by His Fordship, Mustapha Mond, the Great Controller. And Mustapha Mond's power emanated not from him as a person but as a representative of the political *system* of the Brave New World. Thus, I will refer to processes of kind A + B as '*Mustapha Mond sorting*'. I visualize that 'Mustapha Mond sorting' at a given level most often occurs together with selection at that level. I will suggest that this interactive phenomenon is what has been called context-dependent selection at the phenotype level. In interpreting evolutionary phenomena at different levels, we should also consider the possibility of pure 'Mustapha Mond sorting'.

6.2 'Mustapha Mond sorting' among cells in organisms

One way to look at Fig. 1B is to visualize the #'s and ∗'s as variant cells inside organisms (the circles). In metazoan organisms, the sorting among different somatic cellular phenotypes is dictated by their context inside the developing body, as during induction. But this form of dictated sorting inside a higher individual cannot leave an evolutionary imprint—all the sorted body cells are genetically equivalent (hence sorting here refers only to cell phenotypic variation) and, once the germ line has been determined no mutant 'rebel' cells outside it can gain access to the next generation. However, control over ontogenetic differentiation first had to evolve. (Even in many taxa today mutations in developing cell lineages can still gain access to the germ line, Buss 1987). Let us think back to the times when germ line determination was evolving during the beginnings of multicellular organization. 'Mustapha Mond sorting' of heritable cell variants might have occurred when mutants first arose that allowed the metazoan system of some organisms (but not others) to restrict the access of cellular rebels to the next generation. The interplay between selection of such metazoan variants, in opposition and synergism with cellular selection, could have left an imprint on subsequent evolution of development, as Buss (1987) suggested. (It is important not to confuse what Buss 1987 argued in relation to **A**, **B**, and **C** in Fig. 1, taking #'s and *'s to be cells and organisms as circles: he was partly referring to selection among cells which is not represented in Fig. 1 under my assumptions, and partly to the subsequent evolution of systemic [metazoan] control of cell sorting which is represented in **B**, occurring *simultaneously* with metazoan organismal selection, **C**.)

6.3 The debate on group selection

Several models of group selection were reviewed by Wade (1978; see Sober 1984 for a more recent review.) Wade recognized two categories; 'traditional group selection models' (Wright 1945; followed by others); and 'intrademic group selection models' (Wilson 1975, 1977; and others). Most 'traditional' models assumed: (1) that the origin of between-group variation occurred by genetic drift; (2) that groups contribute migrants to a 'migrant pool' from which colonists are drawn at random to fill vacant habitats; and (3) that the mechanism of group selection is differential group extinction. Most 'intrademic' models assumed: (1) that between-group variation originates by sampling error; and all assumed (2) the same role of 'migrant pools' as above; and (3) that group selection operates by differential rates of migration into the common 'migrant pool' (Wade 1978, Table 1).

  I will point out a different distinction that partly overlaps Wade's (1978). It relates to the previous discussion as illustrated in Fig. 1. This fundamental difference between kinds of 'group selection' proposals has also been noted by others (Arnold and Fristrup 1982; Sober 1984; Williams 1985; Maynard Smith 1987; Heisler and Damuth 1987). '*Group selection*' *of type (1)* requires sorting among groups based on heritable differences between them. That is, a 'group offspring' must heritably resemble a *particular* 'group parent' in the character that is subject to sorting. This eliminates all models that assume mixing of genetic elements from different groups in a common 'migrant pool', from which colonists are drawn at random to participate in the generation of new groups. 'Group selection' (1) can be represented by Fig. 1**C**, with the circles representing groups and the #'s and *'s organismal phenotypes. '*Group selection*' *of type (2)* does not require heritable sorting among groups; but only that groups vary in properties that have a causal impact on phenotypic fitnesses. Several of the 'group selection' (2) models can be represented by Fig. 1**B**, others by Fig. 1**A**. Wade's (1977, 1978) own development of group selection based on experiments in *Tribolium* beetles belongs in 'group selection' (1); and a subset of his 'traditional' models seems to as well. All 'intrademic' models are about 'group selection' (2).

In my own concept, group selection is directly analogous to selection among organismal phenotypes and to species selection (see p. 133). **Group selection is the interaction between heritable emergent characters of groups and common environmental elements that causes differences in rates of origination and/or extinction of groups within a species.** The reader will now appreciate why the term 'group selection' for categories (1) and (2) is given in quotation marks: in my view no models of type (2) qualify as group selection; and only some of type (1) do: those requiring that groups vary in stable *group-level* properties (like sex ratios or other compositional or organizational features of the groups themselves), and that each property can be passed on (because its genetic basis is passed on) from an ancestral group to a descendant group.

There is a kind of 'among-group-heritability' implied in models that use the migrant pool assumption. The pooling of phenotypes from different 'parent groups' is in one respect analogous to the mixing of alleles in biparental (sexual) reproduction, except that there is a variable and potentially large number of 'parents' (as pointed out to me by D. S. Wilson and E. Sober personal communication). Such a succession of populations—'parent groups', migrant pools, and colonist groups drawn from migrant pools—can certainly evolve in terms of included genotypes and phenotypes. But the anology with sexually reproducing organisms is limited: The groups in these models have neither the analogue of the selected emergent organismal property (the phenotype), nor the analogue of the 'honest meiotic shuffle' that occurs in sexual organisms (with rare exceptions, Crow 1979). I see the differing probabilities among groups of contributing migrants and of extinction as effects (*sensu* Williams 1966) of context-dependent evolution of genotypes and phenotypes. These models point out interesting evolutionary possibilities that need to be explored. But in my view they are not about group selection. Let us have a closer look at some aspects of 'group selection' (1) and (2).

6.4 'Group selection' of type (1)

Why should heritable sorting among groups, that characterizes 'group selection' (1) processes, be insufficient for group selection? I illustrate this with an example of a cline in shell banding and colour in a snail species of *Cepaea*, that correlates consistently with a north-to-south temperature gradient. Lamotte (1959) found such a correlation. Let us say that local environmental selection of phenotypes is responsible (quite likely; see Lewontin 1974, p. 234). If the temperature increases throughout the area, more of the northern, cold-adapted phenotypes will die. Precisely the same process occurs along a neighbouring north–south transect, where intervening areas of unfavourable habitat happen to have resulted in partly isolated snail groups instead of a continuous cline. Northern groups, because they are composed of cold-adapted phenotypes, differ heritably from southern groups of warm-adapted phenotypes. As temperatures increase we see differential extinction of cold-adapted groups, and founding of new groups by migrants from surviving, warm-adapted groups. In spite of heritable sorting among groups, I would not call this group selection, but effect sorting among groups, because group structure *per se* is unimportant. Phenotypic selection accounts entirely for evolution: it directly effects group sorting along the environmentally-heterogeneous transect, as directly as it results in evolution along the continuous cline.

Wynne-Edwards's (1962) concept, which in the eyes of many initiated the debate on levels of selection, resembles 'group selection' (1), although the role of heritability from 'group parent' to 'group offspring' is unclear in his discussion. He stressed long-term group *survival* as 'the supreme prize in evolution' (1962, p. 20) and did not consider group multiplication in its own right, but only as a secondary consequence of survival. He argued that group selection is a powerful agency of evolution, specifically of the evolution of group adaptation. In this claim he did not rigorously draw the line between properties that emerge at the group level and those that groups have by summation of organismal properties, between properties that merit the name group adaptation and those that represent a direct transfer of organismal adaptation; and therefore between group selection and group sorting as a simply transferred consequence, or effect, of organismal selection. Wynne-Edwards (1962) inspired a large debate, in which a critique of the previous issues featured prominently (Williams 1966; Maynard Smith 1964, 1978). This debate in my view was resolved in favour of the critics.

Elements of Wright's (1931, 1945) early explorations of 'interdemic selection' belong to the 'group selection' (1) category: in a highly subdivided population with limited gene flow some demes, due to the interaction of drift with selection, become fixed for alleles or allelic combinations that confer high absolute fitness. As a result these groups become larger and less subject to further disruption due to drift. They more frequently 'inoculate' other demes with highly fit migrants and give rise to more new demes of high fitness.

In Wade's concept (1977, p. 135) 'group selection is defined as that process of genetic change brought about or maintained by differential extinction and/or proliferation of populations'. This accords with 'group selection' (1) (provided one assumes that he means sorting which is non-random and heritable from 'group parent' to 'group offspring', neither of which is explicit in the definition). The definition is insufficient for group selection in my restricted sense. Yet the design and outcome of some of Wade's (1977) experiments on *Tribolium* beetles can be argued to be consistent with group selection. The experimenter himself imposed group sorting in the true genealogical sense. That is, he did not use a 'migrant pool' but propagated each 'group offspring' directly from a 'group parent'. Furthermore, he sorted groups using a group-level character, namely group size, and the results suggested heritability of this character between groups. Nevertheless, I have difficulty in seeing group size function in nature as an emergent selected character. One can explore a possible natural process, that is similar yet better described as effect sorting among groups in my view: let us say that beetles are distributed in partly isolated groups due to a patchy environment, as in the snail example. Local selective forces differ among patches in variously favouring r- or K-life history variables. That is, some groups contain more fecund (and larger numbers of) beetles than do other groups. There are larger numbers of the fecund phenotypes overall. Assume that each beetle has the same probability of founding a new group, and that habitat patches favouring r- and K-life styles are equally available to migrants. Then fecund phenotypes, which will come more often from large 'parent' groups, will initiate more descendant groups. I have difficulty in seeing this as anything other than group sorting as an effect of phenotypic selection in a heterogeneous environment.

6.5 'Group selection' of type (2)

I suggest that most of what is subsumed under this concept has a strong component of what I call 'Mustapha Mond sorting' (see Fig. 1) acting together with organismal selection, and is crucially different in logical structure to all other concepts of 'X selection', where X has varied from genes to species: 'Group selection' (2) occurs when groups vary in group properties, and those properties have a causal impact on *organismal fitness*. There is no requirement that groups themselves are sorted (although groups differ in probabilities of contributing organisms to the 'migrant pool' (p. 124), this does not amount to group sorting); nor that a selected group property be heritable from a particular 'group parent' to a 'group offspring'. It is required that 'being in a group that has [the group property] is a positive causal factor in the survival and reproduction of organisms' (Sober 1984, p. 314). 'Trait group selection' and 'intrademic group selection' (Wilson 1975, 1977, 1983; Wade 1978; also Heisler and Damuth's 1987 'multilevel selection 1') belong here. In all these versions 'group selection' is 'conceptualized as selection between individuals [here meaning organisms]—but based on an individual's group membership rather than characters attributed to that individual' in Arnold and Fristrup's (1982, p. 117) words.

There is a paradoxical quality to the term 'group selection' in models that focus on the fitnesses and selection of alleles and phenotypes. To be sure, these fitnesses are influenced by group membership, and groups come to differ in composition; but *sorting among* the units of selection (at least those cited in the term group selection) is not required. The chief way in which groups are seen to prosper differentially is by becoming 'fatter' (graphically depicted in Fig. 1A–B)—one might say the issue is group 'fatness', not group fitness.

Later writers on group selection show awareness of this difficulty. Sober (1984), for instance, goes to great length to accommodate it—while he clearly wishes to retain both 'group selection' (1) and (2) in the selection orbit, he realizes that some ingenious conceptual manipulations are required to render the logical differences of 'group selection' (2) from other selection concepts less jarring. He suggests that it is justified, in wrestling with units of selection issues, to *choose a 'benchmark'*—a unit that is the beneficiary of a given selection regime—*located at a different level from that of the property-bearers that are the named units of selection.* Thus, the analysts of 'group selection' (2) were interested in selection and sorting among heritable phenotypes, and so made them the benchmarks. But because group properties (not group sorting) are important to benchmark selection, the term 'group selection' was applied.

Further terminological 'oil on the troubled waters' of the logical discrepancies is proposed by Sober (1984): namely, that we should use the term 'selection *for*' properties of the named units of selection, and 'selection *of*' all objects whose sorting is affected by that selection process. ' "Selection of" pertains to the *effects* of a selection process, whereas "selection for" describes its causes' (Sober 1984, p. 100). This fiat in one stroke makes all 'free riders' hitch-hikers, effect-sorted groups and species, and Mustapha Mond-sorted entities 'selected'. By moving around benchmarks we can stretch the realm of the selected to cover nearly everything. For example, if cells (germ or somatic) vary in having a property that dictates disproportionate multiplication of a non-coding DNA sequence, and if we choose DNA sequences to serve as benchmark, then 'cellular selection' by 'selection for' the property results in 'selection of' the DNA sequence. Is this what is intended? To my mind this scheme is too all-encompassing. I argue that the term selection should be unencumbered by movable benchmarks and prepositional subclassifications, and should be reserved for a kind of process the basic elements of which are consistently the same in different hierarchical contexts.

According to Sober (1984), the difficulties in regard to genic selection are comparable and resolvable by the same means that he applied to group selection: 'selection for' a genic property, with genes the units of selection, and organismic benchmarks. But I disagree that the ongoing gene selection debate and his prescription for group selection have as much in common as he suggests. The gene selection debate concerns the relations between replicators and interactors/vehicles (Dawkins 1976; Hull 1980). These two units of selection (as is now widely accepted; discussed below) are sorted by the same selection regime. The question is: at which level are the interactors, that bear the highest-level heritable properties visible to selection (the aptations), situated? Sober's prescription for 'group selection' (2) aims to include *two levels of selected interactors*, in a selection process 'for' a non-heritable property, and named after the interactor that is not being sorted.

The points made by proponents of 'group selection' (2), for instance by Wilson (1975, 1977, 1983), and those added by Sober (1984), are logically quite clear. I think that the process they see as a form of group selection needs to be considered because of its potential to shape variation patterns within species. Nevertheless, I suggest that we need to tease apart all the various phenomena labelled as 'selection' and retain only those that are consistent with each other. 'Group selection' (2), in all guises that do not involve group heritability and sorting, is not group selection in the sense in which selection is construed in other cases. It should be separated conceptually and terminologically. It does require a group property that is 'emergent'—or whatever one chooses to call those properties at higher levels that are more than simple sums of lower-level properties. This distinction is addressed in the next section.

7. THE CHARACTERS RELEVANT TO SORTING AND SELECTION

7.1 Adaptation, exaptation, and aptation

Darwin noted that characters of organisms, that happen to be useful to them and selected currently, are not necessarily 'adapations' in terms of his own concept: 'The sutures in the skulls of young mammals have been advanced as a beautiful adaptation for aiding parturition, and no doubt they facilitate, or may be indepensable for this act; but as sutures occur in the skulls of young birds and reptiles, which have only to escape from a broken egg, we may infer that this structure has arisen from the laws of growth, and has been taken advantage of in the parturition of the higher animals' (Darwin 1859, p. 197). Williams (1966) took the matter further. In discussing under what circumstances groups can be said to have adaptations, 'biotic adaptations' as he termed them, he argued for a restriction of the term adaptation to such features that have been shaped by a history of selection for a particular function. That is, a character that is currently an adaptation for a function must have been shaped by step-by-step directional selection for that function, and is currently maintained by stabilizing selection for that function.

Building on this foundation, Gould and Vrba (1982) proposed a classification of characters (Table 1) that depends on relative time and history. **Aptations** are all those characters of individuals that are currently subject to selection for a particular function, irrespective of their previous histories. **Adaptations** are those aptations that were shaped by selection for current use. The meaning of 'current' depends on the temporal context in focus, and is as readily applicable to the present as, say, to the Jurassic. For instance, if

Table 1

A taxonomy of characters in relation to selection. Although such characters have usually been discussed at the level of organismal phenotypes, in principle the concepts can apply at levels other than that of organisms. (Adapted from Gould and Vrba 1982.)

A character that is currently selected for a particular function, and was shaped by selection for that function.	Adaptation	
A character that is currently selected with respect to particular role, but that evolved by processes other than selection, or by selection for a different role	Exaptation	Aptation
A character that is currently not subject to selection		Non-aptation

Ostrom's (1979, p. 56) proposal is correct that 'the wing skeleton in *Archaeopteryx* is strong evidence of a prior predatory function of the proto-wing in a cursorial proto-*Archaeopteryx*', then that proto-wing morphology was an adaptation for current predatory function in the Jurassic. An **exaptation** is a character that is currently useful (and subject to selection) in a particular role, but that was not shaped by step-by-step selection for that role. Its previous history may have included no direct selection (such as the skull sutures in Darwin's example, or a mutant molecule that is instantly useful). Or the previous history of an exaptation may have included selection for a different function, with subsequent co-aptation for current use. Non-aptations are characters that are currently not subject to selection. The prominence of the hominine chin may be an example (Gould 1977).

In sum, we argued for the existence of the set of aptations, consisting of partly overlapping subsets of adapations and exaptations, and the set of non-aptations (Gould and Vrba 1982); and for recognition that these concepts in principle apply at several levels of life's organization (Vrba and Gould 1986).

7.2 Aggregate and emergent characters in relation to selection (Table 2)

Among reproducing entities at each level there is variation. They vary in characters arising from summed characters of included, lower-level entities— 'sums-of-parts' or aggregate characters. For instance, organisms may differ in particular non-coding DNA sequences or in the number of body cells; and one species may include only white organisms and another only grey ones.

They also vary in level-specific, 'emergent', structural or dynamic characters. Such characters include phenotypic differences among organisms like the number of forelimb digits; and differences among species in characteristic population composition, and in gene pools. The terms 'level-specific' and

Table 2

*Examples of aggregate and emergent (structural and dynamic) characters of individuals (*sensu *Ghiselin 1974; Hull 1980) that I suggest are not suitable candidates for aptations, and examples of emergent characters that are. The individuals in the table happen to be organisms and sexually reproducing species, but the principles are meant to apply to reproducing individuals at other levels as well.*

	Organisms	Species
Aggregate characters that cannot be aptations	A genome or any part thereof Ecophenotypic characters (non-heritable phenotypic responses to current environment, as part of a reaction norm)	Alleles and phenotypes fixed in species
Emergent characters that cannot be aptations	Any relative frequency pattern among subparts (e.g. among genes, cell types) which is stable in the organism but non-heritable Birth- and death-rates	Any relative frequency among alleles or phenotypes that is non-heritable Speciation and extinction rates Patterns of species and population distribution that are non-heritable
Emergent characters that are candidates for aptation	Heritable phenotypes from proteins to behaviours Heritable norms of reaction	Heritable patterns of species and population distribution Heritable population structure (characteristic size, sex, and age composition) Rate of variation production (by mutation and recombination) Variation patterns of the gene pool itself, provided they can be transmitted to descendant species

(Note: birth- and death-rates cannot be aptations of organisms because this would make the statement of selection tautologous. Yet a population structure, to which birth- and death-rates of organisms have contributed, is a candidate for species aptation provided it is heritable among species.)

'emergent' convey that such characters cannot be said to be properties of any entities at lower levels, but emerge at the focal level. The number of fore limb digits is not a property of a gene or of a cell. The characteristic population composition of breeding groups of impalas (an antelope species, Jarman and Jarman 1974) is not a property of 'impala organisms'; and the species-specific shape of a coral colony is not a property of coral organisms. The term 'structure of an entity' conveys that a particular spatial arrangement among subparts (included entities) is required, and that the disposition of the different kinds of subparts and their relative numbers in this spatial arrangement are important. Furthermore, if arrangements among subparts are evanescent *within* focal entities (or individuals *sensu* Ghiselin 1974)—appearing momentarily by chance only to dissolve into other arrangements—they cannot be referred to as structures of those entities. They are also unimportant to evolution. Not only do structures need to be present consistently (at least at a particular 'ontogenetic' stage and in a particular environment) within individuals, but they must also be able to be passed on faithfully to descendants to be of evolutionary interest. That is, they must be genetically based and heritable.

Reproducing entities also vary in emergent 'dynamic characters' that are potentially relevant to selection. Thus, organisms vary in behaviours and growth rates. Species vary in mutation rates; and all branches in some lineages share a 'variation machine' in the form of sexual reproduction while others do not (Table 2).

I argued (Vrba 1982, 1983, 1984) that sorting by selection requires that variant properties, that interact with the environment to cause sorting, be heritable and emergent at that level. Although heritable emergent properties are necessary for selection, they are not sufficient. For instance, relative frequencies of organismal phenotypes, such as sex ratios or stable polymorphisms, may be passed on from species to species; and have sufficiently long-term heritability to distinguish a clade A from its sister-clade B over millions of years. But if such characters are merely present, without influencing sorting among those lineages, they are not species aptations. They may be termed 'candidates' for species aptation (see Table 2): they emerge at the right level and are heritable; and if in the future they should interact with some new environmental factors to produce non-random sorting, that would be species selection.

Why should we bother with the notion of 'emergence'? After all, it has a dubious reputation. In the past it has been used at times as an excuse for unwarranted anti-reductionism—unwarranted because what emergence means, and what it implies for an evolving system, were not formulated rigorously. But poor practitioners have never robbed a branch of knowledge of its validity (for instance, medical science survived innumerable quacks). Thus, the answer is not to ignore the problem, but to do better. We need to bother with it—to work out a general formulation of emergent properties, and of precisely how they relate to multilevel selection—because the living complexity that all evolutionists want to understand is synonymous with emergent structure and dynamics. If variant reproducing entities, that were 'going it alone' at some distant past times, had not subsequently interacted so as to become subparts of new and more inclusive structures, the tiers of life's hierarchy would never have evolved.

Consider the significance of emergent properties for evolutionary theory in terms of the following. It is realisitc to suppose that, in the origin of each higher level: (1) existing entities first evolved to interact with each other solely by selection at their own level—a situation as yet free of control from the collective context, but in the process of producing it; (2) As a next step, stable emergent properties of *groups* of such entities evolved. 'Downward causation' (in Campbell's 1974 term) from these properties to the fates of lower, included entities ensued—in the form of context-dependent selection and 'Mustapha Mond sorting' at the lower level. These weak forms of control were the harbingers of a more radical form of downward causation from higher emergence; (3) Each higher level became truly launched only once its emergent properties were sufficiently tightly bonded and permanent to initiate higher-level selection.

At any pair of levels (say, genomic constituents and organisms, or single cells and metazoan organisms, or organisms and sexual species) most or all of evolutionary phases (1)–(3) have continued to operate. Thus, in some cases phenotypic selection proceeds independently of conspecific context, in other cases it is markedly group-context-dependent, and the prevalence of other phenotypes may have resulted from species selection superimposed on selection at their own level. A mathematical formulation of how fitnesses and selection at lower and higher levels relate to each other *and to the crucial node of higher emergent properties* seems essential to a general evolutionary theory that encompasses all three processes, and the evolutionary origins of levels. As a start, I suggest that we overcome our reluctance to refer to emergent properties; and instead acknowledge their importance and the need to negotiate them in future research. To repeat Caswell *et al.*'s (1971, p. 39) conclusion, in relation to their modelling of dynamic systems: 'It is an unfortunate fact that emergent properties have sometimes been given an almost mystical character in the literature. They *are* real phenomena, which do arise to confront anyone studying a complex system'.

7.3 Species aptation and species adaptation in relation to species selection

A species aptation is an emergent, heritable species character that is subject to selection. A species adaptation additionally requires that the character was shaped by a history of *species selection* for its current selective role. Some sub-elements of any *species* aptation must be subject to long-term stabilizing or balancing selection at the *phenotype* level. (This is illustrated in examples (3) and (4), section 11 and Table 3.) Otherwise the species character would not have the persistence necessary for species selection to operate. This is not to say that non-selective sorting of phenotypes cannot, in principle, play a crucial role in species selection. For instance, selection for social behavioural phenotypes in all species of taxon A may consistently result in small isolated populations, and therefore in frequent fixation of novel phenotypes by drift and a high incidence of speciation, whereas selection in sister-taxon B produces a low incidence of such fixation by drift and speciation. Thus, random drift plays an integral part in this process of species selection. But the heritable contexts favourable or unfavourable for its occurrence still rely on phenotypic selection. Similarly, a heritable species pattern of small and distantly isolated populations, maintained by selection at the level of dispersal phenotypes, may interact with genetic drift to result in species selection. MacArthur and Wilson (1967) suggested that in some cases one can view a species as an array of populations which are equivalent to 'islands'. They explored the evolutionary consequences of characteristic processes of dispersion, recolonization, and extinction among such populations. Where such characteristics differ heritably between taxa, they are relevant to species selection (see also Wade 1977).

An emergent organismal character, namely a phenotype, has a genetic basis. If it is a multicellular structure, it may also have a complex morphogenetic basis. It develops by interactions among cells that determine cell types and sorting processes among them; and is maintained throughout the organism's life by cellular interactions and cell sorting. Similarly an emergent species character, if it is to have the longevity needed for species selection, must have a genetic basis; although in this case the route from genes to expression at the level of selection is much more circuitous and distant than it is in the case of organismal characters. An emergent species character may be said to 'develop', or be constructed, and maintained by interactions among organisms. The chief constructive sources of such species characters are organismal behaviours, especially of the reproductive and social kinds, but also behaviours and other phenotypes involved in dispersal and migration (or in non-mobility), feeding, habitat selection, and other activities. Therefore, we may expect emergent characters, aptation, and selection at the species level to be most common among the higher animals.

The distinction between adaptation and aptation is of particular relevance to the debate on group and species selection. Many of us, who accepted Williams' (1966) analysis of adaptation, have since been puzzled by the question: Under what circumstances can one visualize the sequential step-by-step evolution of species' *adaptation* by differential speciation and extinction rates, namely by species selection? The requirements for this are very demanding. First of all, very particular kinds of properties are needed; and they must be heritable and vary among lineages. Secondly, for the evolution of adaptation, through generations of species sorting the relevant species characters must be progessively modified. Species selection must act to favour step-by-step directional changes in species structure and dynamics, for the 'function' (species-level) of particular interactions with the environment that make species and lineages 'good speciators' and/or 'evaders of extinction'.

Maynard Smith (1987) discussed this problem. He argued that it is not sufficient that an entity is a unit of selection for it also to be a 'unit of evolution':

> We are asking whether there are entities other than individuals [here meaning organisms] with the properties of multiplication, heredity, and variation, and that therefore evolve adaptations by natural selection. In particular, are genes, or populations, or species, such entities? I shall refer to such entities as 'units of evolution' . . . What makes organisms into units of evolution is that
> i. they have heredity . . . and
> ii. the replicators, or genes, that are responsible for heredity behave in a way that, typically, does not permit within-individual, between-replicator selection. (p. 121)

If condition (ii) is transferred to the species level to mean that there should be no between-replicator selection whatsoever within species for them to be units of evolution, then of course species are never units of evolution. But an expansion of the concept of the 'hereditary basis of the selected phenotype' *to the species level*, surely means the following: There must be constancy within species (that is, no change by selection within species) in *the hereditary basis of the relevant species structure* (or dynamic emergent property). If condition (ii) is construed in this way, then I think that we already have evidence that many species and lineages do satisfy it. (For instance, all 19 extant species of the antelope genus *Cephalophus*, that is monophyletic on morphological grounds and known since the Miocene, share pair-bonding and a non-gregarious, non-migratory population structure. This may be taken as support for the notion that this emergent character of species has been inherited from a common ancestor, which is exactly the reasoning that is commonly applied in cladistic analysis of organismal characters. There are numerous such examples, certainly in the mammalian groups with which I am familiar, e.g. Dorst and Dandelot 1970.) In fact, Maynard Smith (personal communication) does have the expanded concept in mind. It is the appropriate one in analogy with the genome–organism interface. After all, 'within-individual, between replicator selection' is permitted in organisms among 'selfish' non-coding DNA sequences. The point is: they are irrelevant to the organismal phenotype—selection at this level does not 'see' them and can cope with the unseen changes. By analogy, large proportions of the genome *and* the species' pool of phenotypes are irrelevant to the selected emergent property of that species. Species selection neither sees them nor cares whether they change.

Even if species pass tests (i) and (ii) above, there remains the question of species adaptation. The distinction between an aptation (currently subject to selection) and an adaptation (additionally shaped by selection for current use) is helpful here. It seems to me that species aptation may be quite common, especially among higher animals that have complex social behaviour. In contrast, species adaptation, if it exists at all, must be much rarer.

But one can visualize such a phenomenon. Take the maintenance of sex in the face of the cost of meiosis (Williams 1975; Maynard Smith 1978; but see Templeton 1982*a*, p. 98 for the dissenting view that ' "the cost of meiosis" . . . [is] irrelevant in the establishment of a parthenogenetic population from asexual ancestors'). Many now consider it likely that lineage selection is responsible (see Arnold *et al.*, in press). The proposition is that individual selection favours parthenogenesis but opposes sexual reproduction because of the cost of meiosis. However, in the long term, as a consequence of meiosis and recombination, sexual lineages are able to prosper, survive and diversify in spite of environmental change. (I call this lineage selection, and not species selection, because I reserve the term species *sensu stricto*, used without qualification, for sexually reproducing entities, as do many others, Vrba 1985. Thus, the process described is selection favouring lineages of species over lineages of asexual clones, and not selection among species.) If we accept that this process of lineage selection is caused by the interaction of emergent characters (lineage-wide variation patterns) with environments; that still does not mean that lineage selection shaped those characters. So far we have lineage aptations, specifically lineage exaptations. But consider the following: global climatic changes and cycles of differing severities are known to have occurred throughout evolution (review in Vrba 1989), at time intervals varying from tens of thousands, through a few million, to tens of millions of years apart (those causing mass extinctions, possibly with a periodicity of about 26 million years, Raup and Sepkoski 1982). Let us assume correlations, at least in some major phylogenies, between the capacity of reproductive systems to produce high levels of variation and extensive geographic distributions (I know of no studies on this); and between the latter and lineage survival rates through major extinction episodes (Jablonski 1987 provides evidence from marine taxa for this, and also for heritability of breadth of geographic distribution at the lineage level). Then within sexually reproducing phylogenies there may have occurred species selection that progressively favoured systems of more and more efficient variation production.

Calling such a highly elaborated variation 'machinery' common to a set of species a 'species adaptation' is consistent with the recognition that its elements are both currently maintained, and were elaborated, by organismal and genic selection. Turning for a moment to organismal adaptations: before they can be selected, each novel phenotypic element of complex adaptations like the eye must first appear by mutation and via ontogenetic processes that can influence the phenotypic result nonlinearly, filter it or constrain it (Bonner 1982). This does not deter us from acknowledging adaptation and marvelling at the 'creative' power of selection. Similarly, the fact that origin (and maintenance) of a species character has to run an even longer 'ontogenetic gauntlet', that includes the directional forces, filters, and constraints of organismal selection, does not rule out the possibility of stepwise-selected species adaptation.

Still, I think that Williams (1966) and Maynard Smith (1987) are right in arguing that it is an onerous concept. They come to this conclusion without explicitly requiring a species adaptation to be emergent at the species level. The concept is still onerous, but very much less so, if the relevant species characters are interpreted to be emergent at the species level—to argue that certain heritable aspects of population structure or of a species gene pool were put together step-by-step by species sorting is vastly easier than to argue that an organismal phenotype like altruistic behaviour was. (In fact, the latter is extremely unlikely ever to have occurred. Similarly, group adaptations are unlikely to be organismal phenotypes. Even if altruistic phenotypes were promoted by genuine group selection, I would still not agree with Sober 1987, p. 136, that 'altruism . . . is a paradigmatic group adaptation'. I would say that altruistic behaviour is a character of organisms. A characteristic *frequency composition of a group*, of altruistic relative to selfish organisms, may be a group aptation. One can perhaps argue for the existence of an altruistic system among organisms as a group character, and perhaps this is what Sober had in mind.)

8. MODES OF SORTING THAT ARE NOT SELECTION

Before summarizing various views of what selection is, it is useful first to consider what is not selection. Of the five modes of sorting below, probably everyone agrees that most of them are not selection at the sorting level; the exception being those who follow Sober (1984) in applying the term 'selection of' to effect sorting (8.3) and hitch-hiking (8.4). Case 8.5 represents a 'grey' area in the levels-of-selection debate. Several modes of species sorting, as well species selection, are exemplified at the species level in Table 3.

8.1 Random sorting

From its earliest treatments (Wright 1931, 1932, 1945) and including recent discussions (e.g. Crow 1986), random drift has become generally acknowledged as a cause of sorting among phenotypes and genotypes. Random changes in gene frequencies are especially important in small populations (*op. cit.*). An association of drift with population size suggests control by population structure of this mode of gene and phenotype sorting ('downward causation' to sorting, Vrba and Eldredge 1984). But it is recognized to occur in large populations as well (Kimura 1983; Crow 1986). Random sorting can occur at other levels including that of species (Williams 1966; Raup *et al.* 1973). That is, so many different proximal causes act together that the resultant patterns of species diversity have no consistent correlation across clades with any characters of organisms or species. Raup *et al.* (1973) showed that trends, in which phenotypic characters are associated with species diversification, can be simulated if both direction and frequency of speciation events, and frequency of extinction, are allowed to vary randomly.

8.2 Non-heritable sorting

(This kind of sorting at the organism level is also referred to as 'ecophenotypic'.) Birth- and death-rates varying non-heritably as part of a norm of reaction are well known in organisms (Stearns 1982). This occurs facultatively in response to different physical environmental conditions. It is also known to occur in response to other factors. For instance, a salamander that has its tail bitten off by a predator experiences a lower reproductive rate until the tail has regrown (Maiorana 1977). There are probably very few evolutionists (if any) who would call such a temporary and non-heritable response 'selection'. One might be tempted to ask: if the results of this type of sorting are not heritable, then why should it be considered seriously in evolutionary analyses? There are several reasons why non-heritable sorting needs to be studied.

First, evidence suggests that norms of reaction themselves are heritable and subject to selection. It is common that reaction norms, with respect to a particular character, vary within a clade from including a broad range of facultative responses to environmental changes (say, from cold to warm temperatures) in some species, to genetically fixed responses irrespective of environmental change in others. Typically, the fixed differences among species show the same environmental correlations as the facultative ranges within related species (Gould 1977; Wake and Larson 1987). Patterns of this kind have suggested to many biologists that there must be evolutionary mechanisms of converting by natural selection the genetic bases of broad norms of reaction to narrow ones (permanent responses in the limit), and vice versa (see Smallwood's in prep., review of genetic assimilation, which is regarded as a plausible process of the first kind).

Secondly, an analogous process among species needs to be considered in the increasing number of macroevolutionary analyses: Cracraft (1982) suggested that related clades may come to differ non-randomly in speciation and extinction rates, not because they differ in any intrinsic heritable characters, but simply due to their being subjected to different geological and climatic histories. Although he termed such a process 'species selection', I argued that this is the species-level analogue of ecophenotypic sorting (Vrba 1984). A hypothetical example of such sorting among species is given in Table 3, Case 1. Templeton (1986) considers that the exceptional species diversity of Hawaiian drosophilid clades, relative to continental clades, may be a case of what I call non-heritable sorting among speices. That is, the causes derive from the unusual geological history of the Hawaiian islands, rather than from heritable differences between island and mainland clades.

8.3 Effect sorting

The 'effect hypotheses' was originally articulated for a cause of sorting among species that is distinct from species selection, although both processes involve non-random species diversification based on heritable differences among species and lineages (Vrba 1980, 1983). Analogous effect sorting occurs among groups within species, as discussed in section 6.4.

At the species level, the effect hypothesis suggests the following: *heritable characters and selection of organisms,* that are similar within but differ among species and lineages, determine differences in speciation and extinction rates—effect macroevolution or effect species sorting. The important difference from species selection (further discussed below) is that effect sorting involves no causal influence from emergent species structure. The species do differ in characters, but they are aggregate species characters (Table 2). The causal transactions are strictly between each organism and its environment, and the effects flow unopposed upwards to sort species as an incidental byproduct.

The term 'effect' was chosen in recognition of the importance of Williams' (1966) argument: we should not confuse selection (which 'cares' only about immediate fitness of the selected) and adaptation (which is no more than the character shaped directly by selection for current function) with the incidental evolutionary effects of these phenomena. Paterson (personal communication 1978) articulated the idea that organismal selection does not act 'for speciation'; therefore individual speciation events are always effects. Vrba (1980, 1983) argued that effects may extend upwards across hierarchical interfaces—particularly that regimes of organismal selection (differing among the lineages of a monophyletic group) may incidentally yield disparate patterns of species diversification. Biologists have long been accustomed to the idea of species' extinctions as effects: organismal selection, brought on by environmental change, removes the organisms one by one until none is left—species sorting as an effect of organismal selection in the Darwinian sense. The crucial aspect of the effect hypothesis (and its novel component not addressed by previous theory) lies elsewhere: selection for proximal fitness may also, and incidentally, drive *speciation rates.*

8.4 Sorting at a focal level by selection at a higher level

Any selection process among individuals at higher level will inevitably also dictate a component of sorting among included lower entities. (Here we are not referring to the lower entities that form part of the coding and expression 'machinery' of the selected higher character; but to additional lower entities that also reside within the selected individuals.) These lower entities are not themselves selected—they just 'come along for whatever ride' is prescribed by the higher selection regime.

Many examples of this have been documented at the genome–organism interface. For instance, intense selection on a major locus may result in hitch-hiking effects (Maynard Smith 1978). As an example, Templeton (1982b) cites work on human resistance to malaria on the island of Sardinia. Selection for the malaria-resistant phenotype, favouring the X-linked allele that causes glucose-6-phosphate-dehydrogenase deficiency (a malarial adaptation), may entail a hitch-hiking effect at the closely linked locus causing Protan colour blindness. Thus, hitch-hikers experience a systematic effect on their fitnesses by selection acting on other individuals to which they are linked. (Of course, if the lower rider is not associated with the selected character in a non-random way, for instance if a neutral gene assorts independently of the selected gene, then the sorting consequences of any free ride the neutral gene might get are trivial. I suspect that Maynard Smith intended the term hitch-hiker for the linked case.)

The most interesting cases, of lower sorting by higher selection, occur when the lower entities are additionally selected in their own right; and when these selection regimes at different levels act in opposition. Arnold *et al.* (in press) review an example of possible selective conflict between 'selfish' DNA sequences and phenotypes of plethodontid salamanders (after Sessions and Larson 1987; and Roth *et al.* 1988), that seems to have had interesting evolutionary consequences at both levels. Such interlevel phenomena may also have had a much larger significance in the evolution of living diversity. For instance, Buss (1987) explores the consequences that selective antagonisms, occurring between the cellular and organismal levels and during times of metazoan origin, might have had for the astounding range of multicellular ontogenies we see today. Similarly, the *dominance* and diversity (not the origin) of sexual forms, in spite of the cost of meiosis, may reflect positive sorting on a vast scale of sexual phenotypes by lineage selection (Williams 1975; Maynard Smith 1978; Bell 1982; Arnold *et al.*, in press).

8.5 'Mustapha Mond' and context-dependent sorting

Recall that 'Mustapha Mond sorting' refers to that component of sorting at a focal level, that is caused *within* the bodies of higher individuals (Fig. 1A, B), that is not accounted for by focal level selection; but that is controlled by the emergent properties of those higher individuals. Let us think, for example, of organisms as the entities whose sorting is influenced to a greater or lesser extent by the properties of the groups in which they are occurring. Then we can ask: what is the relationship between 'Mustapha Mond sorting', the familiar context-dependent selection among genotypes and phenotypes, and 'group selection' (2) process (such as some called 'trait group' and 'intrademic selection')? Before attempting an answer, I will take a closer look at context-dependent selection.

Various measures of context-dependence of fitnesses and selection of alleles, genotypes, and phenotypes have been proposed as diagnostic of group, and also of species selection (see review in Arnold *et al.*, in press): (1) group membership must account for some of the covariance between fitness and phenotypic traits (Arnold and Fristrup 1982); and (2) group attributes must account for some of the covariance between fitness and phenotypic traits with selection dependent upon group context in which it occurs (Wimsatt 1981; Heisler and Damuth 1987).

I am pessimistic that context-dependence of lower-level fitnesses on its own will get us very far in grappling with higher-level selection. The problem is that every biological process is in some sense context-dependent, from environmental influences on mutation (Shapiro 1983) to selection as an interaction between traits of individuals with the physical and biotic environment. But perhaps it is unfair to bring in the physical environmental context, as the context-dependence under discussion refers to biotic context. So let us consider the latter. In some cases, the context of organisms has been construed to include non-conspecific organisms. For instance, Wilson's trait group is inclusive enough to encompass the notion that 'a leaf miner's trait group is a leaf' (Wilson 1977, p. 158). Because most notions of group context refer to the kinds and numbers of *conspecifics* in the group, let us focus on this. Even here context-dependence is ubiquitous.

At the broad phylogenetic level, comparing asexual and sexual forms, every sexual organism's fitness is heavily dependent on the presence of members of the opposite sex; and within these categories, many young are dependant on the presence of adults, including particular adults. Many people have noted that one can remove large numbers of organisms from a population without any impairment resulting, while 'removing a large part of the body ... [such as] a randomly selected organ from an animal', in other words a functioning structure, has disastrous consequences (Williams 1985, pp. 8–9). This often-cited comparison is not appropriate. The analogue of removing numbers of organisms from a population is removal of numbers of cells from the body; and the analogue of destroying a functioning structure in an animal would more appropriately be destruction of the sexual system by removing all males from the population.

Turning to comparisons within particular species: fitnesses of alleles and phenotypes, *in what is overwhelmingly considered as gene and/or organismal selection*, commonly depend crucially on the kinds and frequencies of other alleles and phenotypes present (for example, effects of inbreeding structure, e.g. Templeton 1982*b*; frequency-dependent selection in mimicry, Wickler 1968, and with respect to 'Evolutionary Stable Strategies', Maynard Smith 1982, 1987; fitness effects that involve the presence of kin, Hamilton 1964*a,b*; and so on). Sober (1987) argues that context-dependence 'can undermine a causal claim' of cause from genic or organismal selection. Others have gone further in the group selection debate, and equate context-dependence of phenotypic fitnesses with higher-level selection. I suggest that, if we consistently take this to its logical conclusion, current evolutionary reasoning will have to undergo massive changes. I doubt that many instances of 'pure' organismal selection will survive, beyond the like of those isolated desert plants, selected in the struggle for existence against the elements, in Darwin's (1859) account.

Thus, context-dependent phenotypic and genotypic selection is ubiquitous, and certainly cannot be claimed to require higher-level selection. One can counter that in many of the examples I cited the boundaries of the 'groups' conferring context-dependence are fuzzy. In contrast, those who proposed measures of context-dependence of fitnesses and selection of alleles, genotypes, and phenotypes as diagnostic of group selection (see review in Arnold *et al.*, in press), were generally thinking of discrete groups. But even in the case of discrete groups, the argument for *diagnosis* does not work. Sober (1984) has comprehensively reviewed why group context-dependence of fitness, for instance as it appears in analyses of variance and covariance, cannot on its own be diagnostic of higher-level selection. This is true whether group selection is construed in my narrow sense (p. 125) or in the broader sense of 'group selection' models which merely require heritable sorting among groups (p. 124). One can cite numerous instances of group-context-dependence of phenotypic and genotypic fitnesses without higher selection; and of no detectable context-dependence in spite of higher selection (see examples in Sober and Lewontin 1982; Sober 1984). I agree with Sober (1984) on these arguments. Some of the examples of species sorting in Table 3 are relevant to this issue.

To return to my earlier question: what is the relationship between 'Mustapha Mond sorting', context-dependent selection among genotypes and phenotypes, and 'group selection' (2) processes including 'trait group' and 'intrademic selection'? Look again at Fig. 1B with #'s and *'s as heritable organismal phenotypes that are sorted non-randomly within groups, the circles; but allow the assumption in the Fig. 1B caption of control of phenotypic sorting by an emergent group property to be either true or not true. Then there are three possibilities of interest in the present discussion:

1. *Only natural selection of phenotypes is operating free of any influence from group context—autonomous phenotypic selection.* Group structure has nothing to do with what happens. This is not difficult to visualize although, as argued above, it is probably less prevalent than generally supposed.

2. *Only 'Mustapha Mond sorting' is operating, without selection at the lower level.* The fact that it is difficult to visualize this gives one an uneasy feeling: has current evolutionary theory succeeded in defining this possibility 'out of existence'? Downward causation from population structure that results in

the fixation of mutants against selection (such as fixation of translocation karyotypes that result in heterozygote disadvantage) might be argued to be pure 'Mustapha Mond sorting'. At other levels one might more readily be able to find examples of this mode. For instance, a non-coding sequence might be induced to mulitply in one unicellular organism by a particular gene, and not in another in which that gene is absent. I believe that we need to retain this category at least as a theoretical possiblity.

3. *'Mustapha Mond sorting' and phenotypic selection are operating together.* In my view this is what is conventionally called context-dependent phenotypic selection. The sorting pattern produced by context-dependent selection is a resultant of two causal forces: one from selection among phenotypes and the other from the emergent properties of the context. **I cannot see a rigorous distinction between this hybrid process ('Mustapha Mond'-plus-selection) at the organismal level, context-dependant selection, and 'group selection' (2) processes in so far they do not include heritable sorting among groups.**

The notion of context-dependent selection covers a wide range of phenomena arranged along a sliding scale—from selection that is minimally to very strongly context-dependent. (An example of the latter is the altruism case, in which altruistic alleles may be favoured among groups although they are at a selective disadvantage within each group.) It is this latter case of strong context-dependence (with a large 'Mustapha Mond' input) that has exercised the imaginations of evolutionary biologists, including mine. I would like to see a logical separation of this phenomenon from other weaker forms of context-dependence. It seems to me that there is something here that is potentially very important to intraspecific evolution, possibly also to promoting speciation (see also Wade 1977), and deserving of distinction. Otherwise why would the school of 'selection' models involving trait groups and demes, that is so blatantly at variance with the Darwinian concept of what selection is, have gained such wide acceptance as something that needs to be distinguished from organismal selection? Nevertheless, I suggest that such a phenomenon should be distinguished under a name other than group selection.

9. SELECTION: A BRIEF SUMMARY REVIEW

I will look briefly at the most important elements that have featured in discussions of selection, and note my impression of their current status in the debate. All of these elements are represented in one way or another in the examples of species sorting in the next section and Table 3.

9.1 The selection statement need not be tautologous

This is now generally agreed upon, after a lengthy debate in which the tautology charge was convincingly countered. (See review in Vrba 1984; also Maynard Smith 1969; Gould 1976; Brady 1979; Sober and Lewontin 1982; Ruse 1982; Byerly 1983.) Past statements of natural selection that referred only to the outcome of differential reproduction and mortality, such as: 'selection ... *is* differential survival and reproduction—and no more' (Futuyma 1979, p. 292), were indeed tautologous and useless for explanatory purposes. As detailed above, there is whole lot of differential survival and reproduction—namely sorting—in nature that neither is, nor is caused by, selection, in terms of almost all definitions. The agreed-upon resolution is as follows. Selection should be formulated as a causal process; and the formulation should refer to *characters* or traits of individuals and to their *interactions with the environment*, such that predictions for sorting result. I bring this old debate up here, because it forced us all to focus on the nature of the characters that actually interact with the environment.

9.2 Units of selection must themselves be sorted on the basis of heritable variation among them

All formulations of genic and organismal selection, of which I am aware, and many of group and species selection, conform with this. The sole exceptions occur in the school of 'group selection' (2) that focuses on how fitnesses of genes and phenotypes evolve within groups (on 'group fatness') by virtue of group structure, and how difference among groups in 'fatness' (but not in group fitness) relate to global lower-level fitness. Because the issue of *heritable* sorting among groups is problematic, and because the proponents of these models wish to use the term *group selection* (with some modification, such as 'trait group' or 'intrademic group selection'), they are forced into the position that units of selection do not need to be subject to heritable sorting.

Darwin (1859) intended natural *selection of organisms* as an explanation of how *organismal traits evolve*, namely for how adaptations evolve (Maynard Smith 1987). He was very much aware that this requires *heritability from parental to offspring organisms*. I suggest that we keep this backbone of the Darwinian selection concept intact as we explore new contexts in which it may apply, with one relaxation: aptations that are exaptations (Table 1) are subject to selection as well (otherwise we cut out, on illogical grounds in my view, a prodigious portion of what is currently viewed as genic and phenotypic selection; see Gould and Vrba 1982). Thus, we should also allow the maintenance and spread of group and species *aptation* as valid domains of group and species selection. Species selection of species aptations can surely result in spectacular evolutionary patterns among lineages; although, unless the aptations are also adaptations, evolution at the species level will remain confined to multiplication, maintenance, and deletion of that which evolved adaptively at lower levels. However, species *adaptations* are not difficult to visualize in principle and probaly exist in at least some cases, although in a form that we have not been inclined to see with our relentless 'organism's eye-view' of adaptations (see pp. 135–7).

9.3 Which are the units of selection?

It seems to me that there has recently been a major swing to the opinion that any one selection process involves *two* distinct operations by units of selection. First, few current evolutionists who recognize that heritability is important for selection, would argue against the position that some genomic particles must always be units of selection as bearers of heritability. Williams (1966) and Dawkins (1976, 1978) were the notable proponents of this view, although these earlier treatments used a restricted notion of the gene not only as the unit of replication, but also as *the* unit of selection. Secondly, during the debates of the past decade, the explicit realization seems to have formed that something that interacts with the environment to bias replication is as close to the core of selection as is replication itself. Even the extreme views of Williams (1966) and Dawkins (1976) have grown in this direction. Williams (1985, p. 3) suggests 'a formal separation of the informational (genetic) aspects [that do the 'bookkeeping'] of natural selection from its concrete ecological aspects [that make 'history']'. Dawkins (1982) proposed the term 'vehicle' for 'any relatively discrete entity, such as an individual organism, which houses replicators, and which can be regarded as a machine pro-grammed to preserve and propagate the replicators that ride inside it' (p. 295). He now regards 'vehicles' as units of selection: 'Replicator survival and vehicle selection are two aspects of the same process' (Dawkins 1982, p. 60).

The earliest unambiguous articulation of two kinds of units of selection that function in selection processes came from Hull (1980, 1988, pp. 408–9): *replicators* that pass on their structures largely intact in successive replica-tions; and *interactors* that interact as cohesive wholes with the environment in such a way that this interaction *causes* replication to be differential. Of Hull's interactors and Dawkins's vehicles, which I see as meaning the same thing, Hull's term is closer to my conception of what occurs during selection: selection is the interaction, between heritable, varying, emergent characters, and the common environmental elements experienced by the variant indivi-duals, that causes differences in birth- and/or death-rates among them. Using this concept, a birth-rate cannot be the emergent character subject to selection, as shown by simple substitution in the above statement. This is why I explicitly excluded birth- and survival-rates from being aptations of organisms (and speciation and extinction rates from being aptations of species) in Table 2.

In sum, a consensus seems to be emerging that *two* sorts of entities function in selection processes: replicators and interactors. In some cases, such as 'selfish' DNA selection (Doolittle and Sapienza 1980), the two functions are located in the same unit of selection. We seem to have reached a stage of recognition that the participation of the genome as a level of selection in any selection process is ineluctable; that interactors of some kind are necessary; and that the distinction between replicators and interactors (or vehicles) is ontological and fundamental to the theory. It remains for us to formulate why we should interpret the interactors to be situated at one level rather than at another in a given selection process (see also Dawkins 1982, p. 413).

9.4 The diagnosis of higher-level selection

The criteria that are sufficient to argue for group or species selection remain in fundamental dispute. Part of the issue has already been generally clarified: William's (1966) argument that a claim of a group selection and group adaptation requires something different from the summed effects of lower-level selection and adaptation, seems to have found widespread consensus. The question is: what is that something different?

Vrba (1980, 1982, 1983, 1984) extended the notion of effect group sorting to the species level by the 'effect hypothesis' of species sorting; and argued that the crucial distinction between what is, and what is not selection at a higher level, lies in *emergent structural or dynamic characters* at that level. Only species selection, by virtue of such characters (aptations), has in principle the power to oppose lower-level selection. Effect sorting of species does not have such power. And the opposition proposed here is not among different selection regimes at the same level (such as selection for large size for escaping predation opposing selection for small size for thermo-regulatory function); but among selection of a higher emergent character that should in principle be able to be antagonistic to selection on phenotypes that are linked to that species structure (Table 3, Case 3; although the antagonism may not be there as in Table 3, Case 4). This is a tall order because any species aptation, in essential outlines, must be based on genotypic/phenotypic variation that is balanced, or at least has a cyclic stability, for long enough for species selection to occur. It is this criterion—that a higher selection regime has a measure of independence from lower ones and, therefore, that it can *in principle* oppose lower-level selection—that led me to suggest that species (and group) aptations require emergent structure and dynamics.

Using the emergence criterion, I propose the following statement of species selection: *species selection is the interaction, between heritable, varying, emergent characters of species and common environmental elements, that causes differences in speciation and/or extinction rates among the varying lineages of species.* The common environmental causes in the case of lineages of species, I suggest, are usually global environmental changes. Species' aptations will most often be those characters of distribution and variation structure that allow species to persist, or that promote lineage splitting, in the face of widespread environmental changes.

I have already outlined why I do not regard patterns of genic, genotypic, or phenotypic fitnesses as sufficient for the diagnosis of species selection. At least part of the problem will be apparent from the hypothetical examples below (Table 3).

10. TESTING FOR EFFECT SORTING VERSUS SELECTION AMONG SPECIES

Several recent studies have explicitly tested for effect sorting versus species selection. For instance, Kitchell *et al.* (1986, working on Late Cretaceous planktonic groups) and Werdelin (1987, comparing eutherian and marsupial groups of carnivorous mammals) concluded that their studies exemplify effect sorting among species. Buss (1988) found that taxa with early ontogenetic determination of the germ line are characterized by low species number relative to taxa with late or variable determination. He concluded that 'this pattern of sorting in higher order taxa . . . [might] be interpreted as an "effect" (*sensu* Vrba 1983), i.e. a case of "upward causation" [using Campbell's 1974 concept]' (Buss 1988, p. 317). In contrast, Jablonski (1987, Late Cretaceous molluscs) and Doolittle (1978, diversification of eukaryote relative to prokaryote lineages due to introns and 'exon shuffling' in the former, but not the latter, genomes) decided that their respective patterns represent species selection and not effect species sorting. I do not agree with all the reasoning variously given in these papers, but cite them here as interesting examples of macroevolutionary patterns and tests of their causes.

In my view, what is needed to test for causes of sorting among species (see, for example, Vrba 1987 for details) includes the following. (A clade is a monophyletic group in the strict sense of Nelson 1971.)

(1) a sufficiently large sample of clades to allow statistical testing of whether any observed correlations are random;

(2) phylogenetic analyses that support hypotheses of monophyly and sister-group status of clades. The characters on which such cladistic analyses are based should be independent of those featuring in the tests of modes of species diversification. Pairwise comparisons of sister-clades are especially favourable;

(3) estimates for speciation and extinction rates (using appropriate corrections, Vrba 1987) for each clade;

(4) independent data on the physical environmental histories, with arguments on common environmental elements that impinged on clades
being compared;

(5) rival hypotheses on how *organismal characters and selection* interacted
with the environment to cause species sorting (effect hypothesis); and on
how emergent species characters might have done so (hypothesis of
species selection), with predictions in each case of associations between
characters and phylogenetic patterns;

(6) testing hypotheses that all such characters are heritable at the level of
species and clades (and testing the alternative hypothesis of non-heritable
sorting among species). A taxonomic sample that has many representatives in the extant biota is particularly favourable for examination of a
wide range of organism- and species-level characters. Such a test for
heritability is based on cladistic analysis;

(7) testing the predictions of the rival hypotheses by comparison of cladistic
distributions of organismal and emergent species characters with speciation and extinction rates.

G. C. Williams (personal communication) is currently of the opinion,
although he recognizes the theoretical distinction between what I call effect
sorting and selection among species, that it is empirically difficult to
distinguish between them. Thus, he includes both processes under 'taxon
selection' (he prefers this term to species selection, Williams 1985), using the
criterion that they share sorting based on heritable differences among taxa.
This broader notion of species selection resembles those of Stanley (1979)
and Arnold and Fristrup (1982). In contrast, I agree with others who argue
that an important distinction is buried in such a definition; that only a part of
what is subsumed under Stanley's (1979) 'species selection' and Williams'
(1985) 'taxon selection' is the logical counterpart at the species level of
natural selection among phenotypes; and that the distinction can be tested in
particular data sets (see Eldredge 1985; Vrba and Gould 1986; Kitchell *et al.*
1986; Gilinsky 1986; Jablonski 1987; Werdelin 1987; Doolittle 1987; Buss
1988; Arnold *et al.* 1989; see also Barnosky's 1987 review). As outlined
above, and as exemplified from my work below, testing which of these two
modes occurred is in my view as feasible as are other widely accepted tests of
long-term evolutionary processes.

On the subject of labelling as selection all higher-level sorting based non-
randomly on heritable differences, I find myself agreeing with Williams'
earlier (1966) rather than later (personal communication) view:

Benefits to groups can arise as statistical summations of the effects of individual
adaptations. When a deer successfully escapes from a bear by running away, we can
attribute its success to a long ancestral period for selection for fleetness ... The same

factor repeated again and again in the herd means not only that it is a herd of fleet deer, but also that it is a fleet herd. The group therefore has a low rate of mortality from bear attack ... As a very general rule ... the fitness of a group will be high as a result of this sort of summation of the adaptations of its members. (Williams 1966, pp. 16–17)

Thus, the summations of the adaptations of members of one deer group may differ from those of another group, resulting in differences between groups in heritable characters and in fitness. Williams (1966) argued that such cases should not be ascribed to adaptation or selection at the group level. Williams's deer example is comparable with the case of *Cepaea* snails (pp. 125–6) that I called effect group sorting. Analogously, I argue that the distinction between effect sorting and selection among species is ontological and important.

10.1 An example based on African mammals (Vrba 1987)

The study was designed to test five rival hypotheses of causes of differences between clades in speciation rates (the 'birth-rate' and 'resource-use hypotheses' [both instances of the effect hypothesis], the 'gene flow hypothesis' [an instance of the species selection hypothesis], and the hypotheses of random and non-heritable sorting). Tests compared modern data on organismal birth-rates, feeding behaviour, water and vegetation cover-dependence, mobility, and population structure, with speciation and extinction rates and trends. Included were 21 monophyletic clades (12 bovid, 2 pig, 1 giraffid, 1 aardvark, 2 elephant, 1 rodent springhare and 2 primate clades). Of these, 14 are hypothesized to constitute 7 sister-groups. All except one are today endemic and abundantly represented in Africa. Almost all were wholly African throughout their known histories, which commenced at various times since the Miocene. I will briefly recount the results of the tests for species selection and effect species sorting.

The 'gene flow hypothesis' (subcase of the *species selection hypothesis*): This states that species are commonly held together by gene flow. Therefore, lineages that are heritably characterized by small isolated and immobile populations should have higher speciation rates than others in which large, genetically cohesive, and mobile populations are prevalent. I tested the prediction of higher speciation rates for clades characterized today by small, sedentary social groups, compared to clades whose extant representatives tend to move about in large migratory groups. The results (see Fig. 2) contradicted this prediction.

The 'resource-use hypothesis' (subcase of the *effect hypothesis*) (*a*) Clades whose resources have tended to persist through environmental extremes during their histories (including generalists or eurytopes, and specialists or stenotypes on resources that range across environments), should have had a low incidence of vicariance (distribution fragmentation) and low speciation

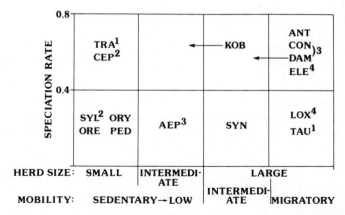

Fig. 2. Herd size and mobility in relation to speciation rate in some mammal clades. Abbreviations in Figs 2–4 are first three letters of taxonomic clade names given in Vrba (1987, tables 1 and 2). For details of how speciation rates were calculated, and sources for data on herd size and mobility in extant representatives, see Vrba (1987). The same numbers are given to sister-clades or to closely related clades. (Adapted from Vrba 1987, fig. 2).

and extinction rates. This is because the organisms experienced relatively little hard selection pressure. Because of the broad range of resource-use and environmental tolerance of each organism, selection was seldom strongly directional and rarely acted to produce major gaps in the continuity of the gene pool (by extirpation).

(*b*) Clades of specialists, whose necessary resources have tended to disappear during recurrent environmental extremes that they encountered during their histories, should generally have had a higher incidence of vicariance, speciation, and extinction. This is because the organisms were subject to relatively frequent episodes of hard selection that wiped out certain phenotypes and populations, and acted strongly directionally on others.

The recurrent environmental extremes to which this hypothesis refers are the 'astronomical' or 'Milankovitch cycles' with global climatic effects with periodicities of approximately 20 000, 40 000, and 100 000 years apart (Berger *et al.* 1984), as well as longer-term climatic changes on the order of one to several million years apart. There is evidence that such climatic changes had major effects on African and other tropical environments (reviews in Vrba *et al.* 1985).

The results of the tests strongly support this hypothesis. Figure 3 shows speciation rate in relation to feeding mode of modern representatives in 18 clades for which the data could be obtained. Figure 4 shows speciation rate in relation to dependence on vegetation cover. In both results, broad resource-use is significantly associated with low speciation rate (and low extinction rates, not shown), and vice versa.

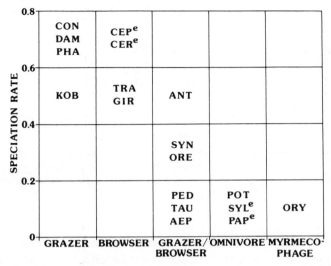

Fig. 3. Feeding categories in relation to speciation rate in some mammal clades. For sources for feeding behaviour see Vrba (1987). e = estimate of speciation rate based on modern diversity. (Adapted from Vrba 1987, fig. 3).

Another important result emerged from this study (Vrba 1987): the magnitudes of divergence trends (long-term net phenotypic change between early and late lineage endpoints) are positively associated with speciation rate. The higher the speciation rate, S, of a clade (in spite of higher extinction rate, E; $S–E > 0$), the more net evolutionary ground is covered at the phenotype level. A hypothetical example of such a correlation is diagramatically shown in Fig. 5. There are at least two ways to interpret this latter result:

1. The selection among specialist organisms promoted higher frequencies and rates of population divergence; and these caused both the larger net phenotypic trend and the higher speciation rates. Viewed in this way, a high speciation rate cannot be said to promote a trend. Both are effects of high tendencies to population divergence.

2. To appreciate the subtle shift in emphasis in the second explanation, recall that 'the species as a field for gene recombination' (in Carson's 1957 term) is sundered in the process of giving rise to a daughter species. Each speciation event represents a partitioning off of a discrete gene pool—a point of no return to future merging with the parental gene pool, because the 'genetic bridges' are severed by reproductive isolation. *In this sense speciation itself has a causal influence on the course of phenotypic evolution.* Speciation most often is an incidental effect of population divergence by organismal selection (Paterson 1978). By its occurrence it provides a feedback causal

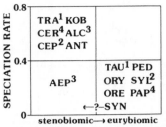

Fig. 4. Dependence on vegetation cover in relation to speciation rate in some mammal clades. The same numbers are given to sister-clades or to closely related clades. Abbreviations and speciation rates as in Fig. 2, after Vrba (1987). (Adapted from Vrba 1987, fig. 4).

influence on the nature of natural selection in the new species. It is not circular to say this. The causal efficacy of speciation arises from its delimitation of a new 'field for gene recombination'. The populations in generalist species, in our example, do evolve by organismal selection; but hard selection less often 'severs the genetic bridges' (results in speciation) than in the case of specialists. At any one time a generalist species tends to be highly polytypic. Through time its phenotypic mean may track climatic oscillations, while its gene pool is likely to retain continuity (unless the climatic extremes exceed the tolerance range of the species, which they are less likely to do in the case of generalists than specialists). Now consider the contrasting situation in specialists specialized on one climatic situation (say warm, moist woodland), as the other climatic extreme (cold, dry grassland) recurs. The model suggests that 'genetic bridges' between parent and daughter populations will be more frequently severed than in generalists. That is, new gene pools (species), each locked into a new 'point of evolutionary departure' in terms of genotypic and phenotypic ranges, will more often arise at one climatic extreme without possibility of return to the parental gene pool once the alternative climate returns. Thus, the long-term equilibration of mean phenotypes that occurs in generalists is precluded in specialists. And every time such new species evolve towards narrower specialization, they have an increased tendency to further irrevocable subdivisions of their gene pools. In this case I think that one can argue that higher speciation rates, in one part of a clade than in another, are promoting a trend. The successive speciation events towards more extremely specialized phenotypes (Fig. 5) act as a ratchet, that contributes to progressively more diverged points of evolutionary departure (and of no return to the ancestral gene pool), and thus to the phenotypic trend.

Figure 5 shows another macroevolutionary effect, that one might predict (albeit somewhat speculatively) if resource-use of organisms relates to environmental changes as postulated. It relates to Gould's (1985) 'three tiers

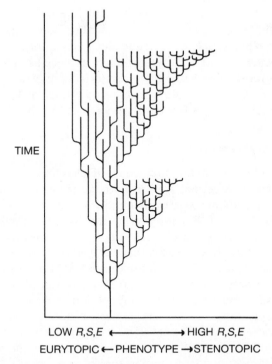

TIME

LOW *R,S,E* ←————————→ HIGH *R,S,E*
EURYTOPIC ← PHENOTYPE →STENOTOPIC

Fig. 5. Iterative trend evolution in a clade. This hypothetical example illustrates the effect hypothesis (that characters and dynamics at the level of organisms can dictate rates of species diversification), and its subcase the resource-use hypothesis (which predicts, simply put, that lineages of eurytopes or generalists should experience low speciation and extinction rates [*S, E*], and vice versa for stenotypes or specialists). The forces driving the trend are operating against a context of 'background macroevolution' interrupted by two mass extinction events.

A complex of linked phenotypic characters varies along the horizontal axis such that phenotypic change towards the right represents increasingly narrow [specialized] resource use. *S, E*, and *R* [net rate of increase in species, $R = S-E$] all increase to the right. The phenotypic trend to the right is clearly not driven by biased speciation direction [here crudely indicated by having as many speciation events to the left as to the right], but by the forces that drive *S*.

of time', the first tier being ecological moments, the second millions of years—the time-scale of 'background macroevolution' on which species selection and effect sorting operate—and the third tier being longer-term episodes of mass extinction. At the first tier of time, selection promotes differing organismal adaptations for resource use. At the second tier specialist species seem to 'inherit the Earth'. At the third tier, it is the specialists that are shaved off wholesale from the Tree of Life by mass extinction. In this scenario, generalists start the cyclic system. Some lineages give rise to specialists. What is selectively equivalent at the organism level (there is nothing 'better' about a generalist than a specialist phenotype; see also census data reported in Vrba 1984), results willy-nilly in disproportion at the species level. At the still more macroscopic level of long-term clades the system is reset, leaving a few hardy generalists to start all over again.

10.2 Additional comments

One can explore a similar interaction of characters, speciation rate, and trends in species selection. And in both cases one can visualize that most species at any one time were evolving towards characters opposite (or quite different) to the trend. For instance, in the example of effect sorting above, evolution within most species may have been towards eurytopy. Rapid speciation of those rare lineages that 'strayed' towards increased stenotypy could still account for the stenotopic trend. In Fig. 5 such independence, between evolution within species and the direction of the trend, is crudely conveyed as follows: there are as many speciation events to the right as to the left phenotypic direction, and species persist in virtual equilibrium for most of their durations, while the net trend is nevertheless towards the right.

Williams (1966, pp. 98–9) makes the same point on the potential independence of within-species from among-species evolution:

From the observation [that Eocene horses are smaller than Pliocene ones] it is tempting to conclude that, at least most of the time and on average, a larger than mean size was an advantage to an individual horse in its reproductive competition with the rest of the population. So the component populations of the Tertiary horse-fauna are presumed to have been evolving larger size most of the time and on the average. It is conceivable, however that precisely the opposite is true. It may be that at any given moment during the Tertiary, most of the horse populations were evolving a smaller size. To account for the trend towards larger size it is merely necessary to make the additional assumption that group selection favoured such a tendency. Thus, while a minority of the populations may have been evolving to larger size, it could have been this minority that gave rise to most of the populations later.

It is interesting that Van Valen (1975) also postulated an antagonism between organismal selection of body size in mammals and 'group selection', but with opposite tendencies to those in Williams' (1966) hypothetical exploration. The 'selected groups' are 'clades' of small body size relative to 'clades' of large body size in the mammalian Tertiary fossil record as a whole. ('Clade' is in quotation marks because Van Valen seems to intend a meaning different from the term as used in cladistic analysis, and elsewhere in this chapter.) Based on the numbers and durations of the 'clades', Van Valen argued as follows: organismal selection consistently promotes evolution to larger size, with small-bodied lineages occasionally giving rise to large-bodied ones but not vice versa. *Ceteris paribus* this should give lineages of large mammals a numerical advantage. But 'small clades' have a 'group selective advantage' (because the net diversification rate is higher among 'small clades' than among 'large clades') that balances the tendency at the organismal level. This is how Van Valen (1975, p. 89) explains his observation in the record of 'two steady states [in numbers of 'clades' per epoch since the Eocene], one of larger mammals and one of smaller mammals'. Unfortunately, he did not specify what he meant by 'clade', 'clade origination and extinction', and precisely how he visualized the operation of 'group selection'. Thus, it is difficult to integrate this fascinating study further in the present discussion.

In sum of this section, at first glance the results of species selection and effect sorting of species may look rather similar (e.g. Fig. 1C with the species represented by circles, and the phenotypes by #'s and *'s, although for species selection an additional species level character would need to be present in the C(ii) species). But there is a crucial distinction, which applies similarly to group selection versus effect group sorting.

In species selection, heritable structures of species, that vary among lineages, interact with the environment to cause species sorting. Such sorting has the power to 'turn around to confront' the lower levels and exert some control over their dynamics. Thus an acid test is this: a sorting process worthy of the name species selection must in principle be able to oppose selection at lower levels (although it need not do so; compare Table 3, Cases 3 and 4).

Effect sorting among species, because it occurs as a direct by-product of summed outcomes of phenotypic selection, or other 'beanbag causes' from lower levels, has no such power to direct patterns at lower levels. It is itself proceeding solely on directions from below.

11. SOME HYPOTHETICAL EXAMPLES OF SORTING AMONG SPECIES

Let us consider five hypothetical examples of species sorting (Table 3). Although they are crude, I put them forward in order to be more specific on the distinctions that were discussed above; and to stimulate discussion. I am asking the reader: do you think that the causal structures of some of these five case histories differ fundamentally from each other; and how would you classify them? I will give my own current views in each case.

The examples represent five different cases of a sister-clade, A–B, of insects. The cases are about differences in fitnesses and absolute numbers of alleles, genotypes and vagile phenotypes (vagile: migratory, freely mobile), and in speciation rates.

The species also vary in two aspects of geographic context: the larger environment surrounding a species may be homogeneous or heterogeneous, and the species' geographic distribution may be continuous or vicariant. (Vicariant means subdivided into isolated population 'islands'.) Before discussing the examples, I need to clarify how I visualize the relationship between these two geographic aspects: Species in which there are no vagile phenotypes have continuous distributions without peripheral vicariance (isolated populations). In contrast, each species with vagile phenotypes thereby gains a particular vicariant distribution structure, with a sprinkling of vagile forms consistently distributed well beyond the peripheries of the main distribution.

A. If the larger environement is *heterogeneous* in topography or substrate, the preferred habitat of the species (containing vagiles) is likely to recur in patches beyond its main distribution. For instance, a species of flies may have its main distribution on a large island with other small islands not too far away; or butterflies may occupy their preferred habitat of a cool upland savanna surrounded by dense lowland forest in which small 'islands' of upland savanna recur due to a patchy topography. In this case the species' vicariant structure is expressed in the form of numerous *permanent isolated populations*, as a result of colonization by vagile phenotypes from the main distribution.

B. On the other hand, if the larger environment surrounding the species is more *homogeneneous*, patches of suitable habitat around the main distribution will be absent. We may think of flies on an island surrounded by huge stretches of ocean; or savanna butterflies that are surrounded by homogeneous rainforests. In this case the species' vicariant structure is expressed in the form of *temporary vicariance*: the heritable species character is maintained by a constant efflux of vagile migrants, but no permanent colonies can take root as suitable habitat is lacking.

I have tried to show this information diagrammatically in Table 3, together with fitnesses and absolute numbers of alleles, genotypes, and phenotypes, and cladistic results. Difference in speciation rate has been arbitrarily indicated in Table 3 by giving one clade two, and the other six terminal branches.

11.1 Case 1 (Table 3): non-heritable sorting among species

As can be seen in Table 4, Case 1, there is no heritable difference between clades A and B. In A and B the same variation in vagility exists within each species: the phenotypes corresponding to the NN and NV genotypes are non-vagile. The rarer VV phenotypes are sufficiently vagile to migrate for considerable distances outside the normal distribution range of the species. The vagiles have lower fitness than the non-vagiles, because their migrations often land them in adverse situations of reduced fitness. The V alleles are nevertheless maintained, because the heterozygote NV has a reproductive advantage over NN and VV. (In this respect some of the present examples are comparable to the case of sickle-cell anemia, e.g. Templeton 1982*b*.) Each species in both clade A and clade B has a vicariant distribution structure by virtue of a low frequency of vagiles. But because A lives in a homogeneous environment, it has a lower rate than B of establishment of peripheral isolates and therefore of speciation. In my view this can neither be termed species selection nor effect sorting, because both require a heritable difference. This is the analogue at the species level of ecophenotypic sorting among organisms.

11.2 Case 2 (Table 3): no sorting among species

Both clades A and B are surrounded by homogeneous environments. The genotypes NN, NV, and VV, that we encountered in Case 1 are present in both A and in B. Yet they differ heritably in the presence of vagile morphs: B has them; but A lacks them because each A-species is fixed for a suppressor gene, X, that is homozygous at another locus. X prevents the VV genotype from being expressed as a vagile phenotype, although it does not suppress the heterozygote advantage of NV. Consequently, clades A and B also differ heritably in the presence of species structure: A is distributed continuously; B has temporary vicariance only. Neither distribution pattern is conducive to speciation; and no sorting among species occurs. Here is a case of species-context-dependent fitnesses and clade-context-dependent fitnesses without higher selection.

11.3 Case 3 (Table 3): species selection in opposition to lower-level selection

Clades A and B are surrounded by the same kinds of heterogeneous environments. Again the genotypes NN, NV, and VV are present in A and in B. As in Case 2, clades A and B differ heritably in the presence of vagile phenotypes (as before, B has them; but A lacks them because species are fixed for the suppressor gene X). Based on the genetic and phenotypic differences, A and B also differ heritably in their species-level distributions. As a consequence A remains at low species-diversity, while B speciates rapidly. I call this species selection. We see here an example of antagonism between selection regimes at genotypic/phenotypic and species levels: genotype VV is of lower fitness than NN and NV in clade B; but in Clade A VV and NN are equally fit, and both less fit than NV. Yet in the end VV does better in B than in A, by virtue of B's high speciation rate. Alleles V and N are both less fit within each species of B than they are within each species of A. Yet in the end V and N do better in B than in A, by virtue of B's high speciation rate. At the phenotypic level vagiles are at a selective disadvantage; yet at the species level the structures vagiles give rise to are positively selected.

Table 3

*Hypothetical examples of species sorting**

Clade	Larger environment surrounding the species	Allele/genotype	Fitness	Phenotype	No. alleles/genotypes per species	per clade	Causal distribution structure per sp. and resulting phylogeny
Case 1: non-heritable sorting among species							
A (2 spp.)	Homogeneous	N	0.91		1780	3560	Temporarily vicariated
		V	0.91		220	440	
		NN	0.90	Non-vagile	792	1584	
		NV	1.00	Non-vagile	196	392	
		VV	0.20	Vagile	12	24	
B (6 spp.)	Heterogeneous	N	0.91		1780	10680	
		V	0.91		220	1320	
		NN	0.90	Non-vagile	792	4752	
		NV	1.00	Non-vagile	196	1176	
		VV	0.20	Vagile	12	72	Vicariated
Case 2: no sorting among species							
A (2 spp.)	Homogeneous	N	0.95		1000	2000	Continuous
		V	0.95		1000	2000	
		NN–XX	0.90	Non-vagile	250	500	
		NV–XX	1.00	Non-vagile	500	1000	
		VV–XX	0.90	Non-vagile	250	500	
B (2 spp.)	Homogeneous	N	0.91		1780	3560	Temporarily vicariated
		V	0.91		220	440	
		NN	0.90	Non-vagile	792	1584	
		NV	1.00	Non-vagile	196	392	
		VV	0.20	Vagile	12	24	

Case 3: species selection

A (2 spp.)	Heterogeneous						
		N	0.90		1000	2000	Continuous
		V	0.90		1000	2000	
		NN–XX	0.80	Non-vagile	250	500	
		NV–XX	1.00	Non-vagile	500	1000	
		VV–XX	0.80	Non-vagile	250	500	
							Vicariated

B (6 spp.)	Heterogeneous						
		N	0.88		1200	7200	
		V	0.88		800	4800	
		NN	0.80	Non-vagile	360	2160	
		NV	1.00	Non-vagile	480	2880	
		VV	0.70	Vagile	160	960	

Case 4: species selection

A (2 spp.)	Heterogeneous	NN Fixed	1.0	Non-vagile	1000	2000	Continuous
B (6 spp.)	Heterogeneous	VV Fixed	1.0	Vagile	1000	6000	
							Vicariated

Case 5: effect sorting among species

A (6 spp.)	Heterogeneous	NN Fixed	1.0	Non-vagile	1000	6000	Distribution structure is not causally involved in species sorting
		SS Fixed	1.0	Specialist	1000	6000	
B (2 spp.)	Heterogeneous	VV Fixed	1.0	Vagile	1000	2000	
		GG Fixed	1.0	Generalist	1000	2000	

*Five different cases of insect sister-clades, A and B, compared with respect to evolutionary environments, speciation rates, presence of alleles N and V (on which are based non-vagile and vagile phenotypes; heterozygote NV has reproductive advantage), and S and G (on which are based habitat specialist and generalist phenotypes). X is a suppressor gene homozygous at another locus and fixed across some of the species which prevents expression of the vagile phenotype of VV. Each species contains 1000 organisms. (See text for discussion)

11.4 Case 4 (Table 3): species selection in synergism with lower-level selection

Clades A and B are surrounded by similar—heterogeneous—environments. They differ heritably in the phenotypic and species-structural characters that are relevant to species sorting. But this time, the ancestral stem of A became fixed for N. In clade B selection for feeding on a new food resource favoured V sufficiently to outweigh negative selection due to migrant losses. V became fixed in the ancestral species of clade B. Again I would call this species selection. But how do we perform analyses of context-dependence here? (See also discussion in Sober 1984 on the 'absent-value problem'.) All genes and phenotypes that characterize B relative to A hitch-hike on species selection. There is no obvious antagonism: organismal selection promotes an adaptation, a feeding behaviour, based on vagility. Thus, V and vagility are positively selected within species. In addition, their representation is vastly boosted by an independent selection regime at the species level—independent in the sense that species selection is blind to how the organisms in A and B feed, and promotes B solely for its species aptation of distribution structure.

11.5 Case 5 (Table 3): effect sorting among species

As in the last case, clade A contains only N alleles and B only V, and they are both surrounded by heterogeneous environments. But let us introduce some additional elements: all the NN non-vagiles in A are also homozygous for S and therefore feeding specialists. The VV vagiles are homozygous for G and feeding generalists. Let us further suppose that the entire sister-group evolves in an area that responds strongly to climatic oscillations. Although clade B does have heritable population structure as a result of vagility, it never becomes relevant to species sorting in this case: the organisms in B species are so generalized—able to eat whatever they find and cope with a wide range of environmental alteration—that they simply do not experience sufficient directional selection pressure, and their populations seldom diverge to the point of speciation. In contrast, the specialists in A experience strong selection pressure as their resources are recurrently withdrawn by climatic extremes. Their populations diverge frequently, and speciation rate is enhanced relative to B. In this scenario the individual responses of organisms to selection, and the summed consequences of lineage divergence that result, are all-important. Heritable species-level structure is causally inert and incidental. Thus I would call this effect sorting of species.

12. SUMMARY

The discussion of levels of selection is currently experiencing a renaissance. Some suspect, as I do, that selection at multiple levels has contributed far more to evolutionary patterns than was recognized in the past. Others are more sceptical. They nevertheless seem to agree that, as important causes *may* indeed flow from processes at and between levels, beyond those traditionally addressed, we had better at least investigate the possibility.

For study of processes at multiple levels, the traditional conceptual framework needs to be enlarged. The causal interactions between the level of coding genes and the level of the organism are much more tightly constrained than are those between any other two levels of the genealogical hierarchy. In the kinds of organisms on which most evolutionary theory is based, the 'bottleneck life cycles' prevent any mutation and changes among mutant somatic cell lineages from affecting the next generation. Furthermore, with rare exceptions, the discipline of meiosis prevents any within-organism selection among genes (where 'genes' is restricted to genomic elements with phenotypic effects). This differs greatly from the relations across the organism-group and organism-species interfaces, and also from what we know to date of the interactions between the non-coding and coding parts of the genome (e.g. Dover and Flavell 1982; Doolittle 1987).

Thus, in order to assess the importance of multilevel selection, the meanings of 'selection', 'adaptation', and of other familiar terms, need to be delimited precisely, set in context alongside additional concepts with alternative or more inclusive meanings; and transferred to other levels only if consistently possible. We need a sufficiently rich conceptual structure to pinpoint such equivalences as do exist between levels, and to distinguish not only the gross but also the subtle differences between levels in so far as they have bearing on evolution. This chapter is a modest effort to add to the contributions that others have already made to an expanded conceptual framework.

The notion of sorting helps to set selection in context. Sorting is a simple description of differential birth and death processes among reproducing entities at whatever level they may be situated. It contains in itself no statement about causes. Darwinism has provided a theory for one cause of sorting—natural selection acting upon heritable charaters of organisms. However, a number of other processes produce sorting as well.

As a general statement of selection, that applies at all levels, I propose: selection is the interaction, between heritable, varying, emergent characters, and the common environmental elements experienced by variant individuals, that causes differences in birth and/or death rates among them. The term individual is here used in the sense of Ghiselin (1974) and Hull (1980). For a process of species sorting to qualify as species selection, it must have a measure of independence from selection among organisms. One must be able to argue that it can *in principle* oppose lower-level selection, although in a given comparison it may happen not to do so. It is this criterion that led me to suggest that species and group aptations (that is, species and group characters subject to selection) require *emergent* structure or dynamics.

A consensus seems to be emerging that *two* sorts of entities function in each selection process: replicators and interactors (Hull 1980, 1988; 'interactors' are equivalent to Dawkins' 1982 'vehicles'). The interactors bear the emergent character variation, and the replicators are responsible for character heritability in the statement of selection above. In some cases, such as 'selfish' DNA selection (Doolittle and Sapienza 1980), the two functions are located in the same unit of selection. We seem to have reached a stage of recognition that the participation of the genome as a level of selection in any selection process is ineluctable; that interactors of some kind are as necessary; and that the distinction between replicators and interactors (or vehicles) is ontological and fundamental to the theory (see also Dawkins 1982). It remains for us to formulate why we should interpret the interactors to be situated at one level rather than at another in a given selection process.

Besides selection, several other modes of sorting may be distinguished at different levels: random sorting, ecophenotypic (non-heritable) sorting, effect sorting (by effects from lower-level selection), and sorting by selection at a higher level (hitch-hiking, Maynard Smith 1978). There is one other kind of sorting that should theoretically be distinguished: sorting among lower entities (such as organisms) can in principle occur *within* the bodies of higher entities (such as groups), dictated entirely by the higher entities' emergent characters. In this chapter I termed the process 'Mustapha Mond sorting'. I have explored the relationship between 'Mustapha Mond sorting' among organisms and what has been termed 'trait group selection' and 'intrademic group selection' (Wilson 1975, 1977, 1983; Wade 1978). Most of these models of 'group selection' address group prospering only in terms of fitnesses of included organisms (group 'fatness') and not in terms of group-more-making (group fitness). I suggest that such processes, in so far they involved no heritable sorting *among* groups, are conceptually separate from group selection (in analogy with what selection means at other levels), and should be given another name. My current conclusion is that context-dependent (meaning conspecific context) organismal selection cannot be rigorously distinguished from 'group selection' models such as for 'trait group' and 'intrademic selection' (in so far as no heritable group sorting is required).

I suspect that species selection will be found to be quite common, once data on true species-level (emergent) characters become available. Species adaptation is a very much more difficult phenomenon to visualize than is species aptation, specifically species exaptation. I have suggested some examples, to stimulate discussion, of species structural and dynamic characters. The debate on species selection has been hampered by our inability to expand the concept of an organismal phenotype to the analogous phenomenon at the species level. Species selection cannot shape an organismal adaptation; but it may on occasion shape a heritable distributional character or a variation structure at the species level. Most often, when species selection occurred, it probably only sorted species exaptations.

Effect sorting among species may well have been a common occurrence in the history of life. Both species selection and effect sorting of species imply an explanation of long-term trends and of other species-diversity patterns that is ontologically distinct from the traditional explanations. To stimulate discussion with my colleagues, I explored hypothetical examples of several different modes of species sorting, in terms of differential speciation rates, environmental contexts, heritability and emergence of characters, and numbers of alleles, genotypes and phenotypes.

Finally, it will be apparent from my discussion that I doubt the validity, in much of the units of selection debate, of pitting 'reductionists' versus 'the rest' (e.g. Wimsatt 1980; Williams 1985). The important distinction seems to me to be between a more simple or direct reduction versus a more complex or indirect reduction of a higher-level phenomenon. For instance, in explaining an among-species trend, one might take the simple reductionist route of regarding gene frequency change by gene selection as a sufficient explanation. Or one might postulate that reference is needed to dynamics among genes, to ontogenies of organisms, to organization and sorting of species, and to the relevant connections between them. The second approach is also reductionist, but it involves a more complex reduction.

ACKNOWLEDGEMENTS

I have received many useful comments from Leo Buss, John Gatesy, Tim Goldsmith, David Hull, Allan Larson, John Maynard Smith, Mark Norell, Alan Templeton, Rimas Vaisnys, and George Williams: I am grateful to all of them.

REFERENCES

Arnold, A. J. and Fristrup, K. (1982). The theory of evolution by natural selection: a hierarchical expansion. *Paleobiology* **8,** 113–29.

Arnold, S. J. *et al.* (1989). How do complex organisms evolve? In *Complex organismal functions: integration and evolution in vertebrates.* (ed. D. B. Wake and G. Roth) Dahlem Konferenzen. Wiley, Chichester.

Barnosky, A. D. (1987). Punctuated equilibrium and phyletic gradualism. In *Current mammalogy,* Vol. 1 (ed. H. H. Genoways), pp. 109–47. Plenum Press, New York.

Bell, G. (1982). *The masterpiece of nature: the evolution and genetics of sexuality.* University of California Press, Berkeley.

Berger, A., Imbrie, J., Hays, J., Kukla, G., and Saltzman, B. (ed.) (1984). *Milankovitch and climate, Parts 1 and 2.* Reidel, Dordrecht, Boston, Lancaster.

164   Elisabeth S. Vrba

Bonner, J. T. (ed.) (1982). *Evolution and development*. Springer, Berlin, Heidelberg, New York.

Brady, R. H. (1979). Natural selection and the criteria by which a theory is judged. *Syst. Zoo.* **28**, 600–21.

Brandon, R. (1978). Adaptation and evolutionary theory. *Studies Hist. Phil. Sci.* **9**, 181–206.

Brookfield, J. F. Y. (1986). A model for DNA sequence evolution within transposable element families. *Genetics* **112**, 393–407.

Buss, L. W. (1987). *The evolution of individuality*. Princeton University Press, Princeton, New Jersey.

—— (1988). Diversification and germ-line determination. *Paleobiology* **14**, 313–21.

Byerly, H. C. (1983). Natural selection as a law: Principles and processes. *Amer. Natur.* **121**, 739–45.

Campbell, T. (1974). 'Downward causation' in hierarchically organized biological systems. In *Studies in the philosophy of biology*, (ed. F. J. Ayala and T. Dobzhansky), pp. 179–86. University of California Press, San Francisco.

Carson, L. H. (1957). The species as a field for gene recombination. In *The species problem* (ed. E. Mayr), pp. 23–38, Publication No. 50. American Association for the Advancement of Science, Washington.

Caswell, H., Koenig, H. E., Resh, J. A., and Ross, Q. E. (1972). An introduction to systems science. In *Systems analysis and simulation in ecology*, Vol. II (ed. B. C. Patten), pp. 3–78. Academic Press, New York.

Cracraft, J. (1982). A non-equilibrium theory for the rate-control of speciation and extinction and the origin of macroevolutionary patterns. *Syst. Zool.* **31**, 348–65.

Crow, J. F. (1979). Genes that violate Mendel's rules. *Sci. Amer.* **240**, 134–46.

—— (1986). *Basic concepts in population, quantitative, and evolutionary genetics*. W. H. Freeman, New York.

Darwin, C. (1859). *On the origin of species by means of natural selection, or the preservation of favoured races in the struggle for life*. John Murray, London.

Dawkins, R. (1976). *The selfish gene*. Oxford University Press.

—— (1978). Replicator selection and the extended phenotype. *Zeit. für Tierpsychol.* **47**, 61–76.

—— (1982). *The extended phenotype*. Oxford University Press.

Doolittle, W. F. (1987). The origin and function of intervening sequences: a review. *Amer. Natur.* **130**, 155–85.

—— and Sapienza, C. (1980). Selfish genes, the phenotype paradigm and genome evolution. *Nature* **284**, 601–3.

Dorst, J. and Dandelot, P. (1970). *A field guide to the larger mammals of Africa*. Collins, London.

Dover, G. A. and Flavell, R. B. (ed.) (1982). *Genome evolution*. Academic Press, London.

Eldredge, N. (1985). *Unfinished synthesis*. Oxford University Press.

—— and Gould, S. J. (1972). Punctuated equilibria: An alternative to phyletic gradualism. In *Models in paleobiology* (ed. T. J. M. Schopf), pp. 82–115. Freeman, Cooper and Co., San Francisco.

Falconer, D. S. (1981). *Introduction to quantitative genetics*, (2nd edn). Longman, London.

Futuyma, D. J. (1979). *Evolutionary biology*. Sinauer Associates, Sunderland, Massachusetts.

Ghiselin, M. T. (1974). A radical solution to the species problem. *Syst. Zool.* **25**, 536–44.

Gilinsky, N. L. (1986). Species selection as a causal process. *Evol. Biol.* **20**, 249–73.

Gould, S. J. (1976). Darwin's untimely burial. *Nat. Hist.* **85**, 24–30.

—— (1977). *Ontogeny and phylogeny*. Belknap Press, Cambridge, Massachusetts.

—— (1985). The paradox of the first tier: an agenda for paleobiology. *Paleobiology* **11**, 2–12.

—— and Vrba, E. S. (1982). Exaptation—a missing term in the science of form. *Paleobiology* **8**, 4–15.

Haldane, J. B. S. (1954). The measurement of natural selection. *Proceedings of the IXth International Congress on Genetics* **1**, 480–7.

Hamilton, W. D. (1964a). The genetical evolution of social behaviour. I. *J. Theor. Biol.* **7**, 1–16.

—— (1964b). The genetical evolution of social behaviour. II. *J. Theor. Biol.* **7**, 17–32.

Heisler, I. L. and Damuth, J. (1987). A method for analyzing selection in hierarchically structured populations. *Amer. Natur.* **130**, 582–602.

Hull, D. S. (1980). Individuality and selection. *Ann. Rev. Ecol. Syst.* **11**, 311–32.

—— (1988). *Science as a process: an evolutionary account of the social and conceptual development of science*. University of Chicago Press, Chicago.

Huxley, A. (1946). *Brave new world*. Harper & Row, New York.

Jablonski, D. (1987). Heritability at the species-level: analysis of geographic ranges of cretaceous mollusks. *Science* **238**, 360–3.

Jarman, P. J. and Jarman, M. V. (1974). Impala behaviour and its relevance to management. In *The behaviour of ungulates and its relation to management* (ed. V. Geist and F. Walther), pp. 871–81. International Union for Conservation of Natural Resources, Morges, Switzerland.

Kimura, M. (1983). *The neutral theory of molecular evolution*. Cambridge University Press.

Kitchell, J. A., Clark, D. L., and Gombos, A. M. (1986). The selectivity of mass extinction: causal dependency between life history and survivorship. *Palios* **1**, 504–11.

Lamotte, M. (1959). Polymorphism of natural populations of *Cepaea nemoralis*. *Cold Spring Harbor Symposium Quantitative Biology* **24**, 65–86.

Lande, R. and Arnold, S. J. (1983). The measurement of selection on correlated characters. *Evolution* **37**, 1210–26.

Lewontin, R. C. (1970). The units of selection. *Ann. Rev. Ecol. Syst.* **1**, 1–16.

—— (1974). *The genetic basis of evolutionary change*. Columbia University Press, New York.

MacArthur, R. H. and Wilson, E. O. (1967). *The theory of island biogeography*. Princeton University Press, Princeton, New Jersey.

Maiorana, V. C. (1977). Tail autotomy, functional conflicts and their resolution by a salamander. *Nature* **265**, 533–5.

Maynard Smith, J. M. (1964). Group selection and kin selection. *Nature* **201**, 1145–7.

—— (1969). The status of neo-Darwinism. In *Towards a theoretical biology* (ed. C. H. Waddington), pp. 35–43. Aldine, New York.

—— (1976a). Group selection. *Q. Rev. Biol.* **51**, 277–83.

—— (1976b). A short term advantage for sex and recombination through sib-competition. *J. Theor. Biol.* **63**, 245–58.

—— (1978). *The evolution of sex.* Cambridge University Press.

—— (1982). *Evolution and the theory of games.* Cambridge University Press.

—— (1987). How to model evolution. In *The latest on the Best* (ed. J. Dupré), pp. 119–31. A Bradford Book, MIT Press, Cambridge, Massachusetts.

Mayr, E. (1978). Evolution. In *The evolution of sex*, pp. 2–11. W. H. Freeman, San Francisco.

—— (1988). *Towards a new philosophy of biology.* Belknap, Harvard, Massachusetts.

Nelson, G. J. (1971). Paraphyly and polyphyly: redefinitions. *Syst. Zool.* **20**, 471–2.

Orlove, M. J. (1975). A model of kin selection not invoking coefficients of relationship. *J. Theor. Biol.* **49**, 289–310.

Ostrom, J. H. (1979). Bird flight: how did it begin? *Amer. Sci.* **67**, 46–56.

Paterson, H. E. H. (1978). More evidence against speciation by reinforcement. *S. Afr. J. Sci.* **74**, 369–71.

Raup, D. M. and Sepkoski, J. J., Jr (1982). Mass extinctions in the marine fossil record. *Science* **215**, 1501–3.

——, Gould, S. J., Schopf, T. J. M., and Simberloff, D. (1973). Stochastic models of phylogeny and the evolution of diversity. *J. Geol.* **81**, 525–42.

Roth, G., Rottluff, B., and Linke, R. (1988). Miniaturization, genome size and the origin of functional constraints in the visual system of salamanders. *Naturewiss.* **75**, 297–304.

Roux, W. (1881). *Der Kampf der Teile im Organismus.* W. Engelmann, Leipzig.

Ruse, M. (1982). *Darwinism defended.* Addison-Wesley, Reading, Massachusetts.

Sessions, S. K. and Larson, A. (1987). Developmental correlates of genome size in plethodontid salamanders and their implication for genome evolution. *Evolution* **41**, 1239–51.

Shapiro, J. A. (1983). Variation as a genetic engineering process. In *Evolution from molecules to men* (ed. D. S. Bendall), pp. 253–72. Cambridge University Press.

Smallwood, P. D. (in prep.). The use and disuse of genetic assimilation.

Sober, E. (1984). *The nature of selection, evolutionary theory in philosophical focus.* MIT Press, Cambridge, Massachusetts.

—— (1987). Comments on Maynard Smith's 'How to model evolution'. In *The latest on the best* (ed. J. Dupré), pp. 133–45. A Bradford Book, MIT Press, Cambridge, Massachusetts.

—— and Lewontin, R. C. (1982). Artifact, cause and genic selection. *Philos. Sci.* **49**, 157–80.

Stanley, S. M. (1979). *Macroevolution: pattern and process.* W. H. Freeman, San Francisco.

Stearns, S. C. (1982). The role of development in the evolution of life histories. In *Evolution and development* (ed. J. T. Bonner), pp. 237–58. Springer, Berlin.

Templeton, A. R. (1979). A frequency-dependent model of brood selection. *Amer. Natur.* **114**, 515–24.

—— (1982a). The prophecies of parthenogenesis. In *Evolution and genetics of life histories* (ed. H. Dingle and J. P. Hegmann), pp. 75–101. Springer, New York.

—— (1982b). Adaptation and the integration of evolutionary forces. In *Perspectives on evolution* (ed. R. Milkman), pp. 15–31. Sinauer Associates, Sunderland, Massachusetts.

—— (1986). The relation between speciation mechanisms and macroevolutionary patterns. In *Evolutionary processes and theory*, (ed. S. Karlin and E. Nevo), pp. 497–512. Academic Press, New York.

Van Valen, L. (1975). Group selection, sex, and fossils. *Evolution* **29**, 87–94.

Vrba, E. S. (1980). Evolution, species and fossils: how does life evolve? *S. Afr. J. Sci.* **76**, 61–84.

—— (1982). The evolution of trends. In *Contribution to Colloque Internationaux du Centre National de la Recherche Scientifique*, (ed. J. Chaline), May 1982, Dijon, France.

—— (1983). Macroevolutionary trends: new perspectives on the roles of adaptation and incidental effect. *Science* **221**, 387–9.

—— (1984). What is species selection? *Syst. Zool.* **33**, 318–28.

—— (1985). Introductory comments on species and speciation. In *Species and speciation* (ed. E. S. Vrba), pp. ix–xviii). Transvaal Museum Monograph No. 4, Pretoria.

—— (1987). Ecology in relation to speciation rates: some case histories of Miocene-Recent mammal clades. *Evol. Ecol.* **1**, 283–300.

—— (1989). What are the biotic hierarchies of integration and linkage? In *Complex organismal functions: integration and evolution in vertebrates*. (ed. D. B. Wake and G. Roth) Dahlem Konferenzen. Wiley, Chichester.

—— and Eldredge, N. (1984). Individuals, hierarchies and processes: towards a more complete evolutionary theory. *Paleobiology* **10**, 146–71.

—— and Gould, S. J. (1986). The hierarchical expansion of sorting and selection: sorting and selection cannot be equated. *Paleobiology* **12**, 217–28.

——, Burckle, L. H., Denton, G. H., and Partridge, T. C. (ed.) (1985). Palaeoclimate and evolution I. *S. Afr. J. Sci.* **81**, 224–75.

Wade, M. J. (1977). An experimental study of group selection. *Evolution* **31**, 134–53.

—— (1978). A critical review of the models of group selection. *Q. Rev. Biol.* **53**, 101–14.

Wake, D. B. and Larson, A. (1987). Multidimensional analysis of an evolving lineage. *Science*, **238**, 42–48.

Weismann, A. (1904). *The evolution theory*. Edward Arnold, London.

Werdelin, L. (1987). Jaw geometry and molar morphology in marsupial carnivores: analysis of a constraint and its macroevolutionary consequences. *Paleobiology* **13**, 342–50.

Wickler, W. (1968). *Mimicry*. Weidenfeld & Nicolson, London.

Williams, G. C. (1966). *Adaptation and natural selection*. Princeton University Press, Princeton, New Jersey.

—— (1975). *Sex and evolution*. Princeton University Press, Princeton, New Jersey.

—— (1985). A defense of reductionism in evolutionary biology. In *Oxford surveys in evolutionary biology*, Vol. 2 (ed. R. Dawkins and M. Ridley), pp. 1–27. Oxford University Press.

Wilson, D. S. (1975). A theory of group selection. *Proc. Natl Acad. Sci. USA* **72**, 143–6.

—— (1977). Structured demes and the evolution of group-advantageous traits. *Amer. Natur.* **111**, 157–85.

—— (1983). The group selection controversy: history and current status. *Ann. Rev. Ecol. Syst.* **14**, 159–87.

Wimsatt, W. C. (1980). Reductionist research strategies and their biases in the units of selection controversy. In *Scientific discovery: case studies* (ed. T. Nickles), pp. 213–59. Reidel, Boston.

—— (1981). The units of selection and the structure of the multi-level genome. In *PSA 1980, Vol. 2* (ed. P. Asguith and R. Giere). Philosophical Science Association. East Lansing, Michigan.

Wright, S. (1931). Evolution in Mendelian populations. *Genetics* **16**, 97–159.

—— (1932). The roles of mutation, inbreeding, crossbreeding, and a selection in evolution. *Proceedings of the VIth International Congress on Genetics* **1**, 356–66.

—— (1945). Tempo and mode in evolution: a critical review. *Ecology* **26**, 415–19.

Wynne-Edwards, V. C. (1962). *Animal dispersion in relation to social behavior.* Oliver & Boyd, Edinburgh.

Yokoyama, S. and Templeton, A. (1980). The effect of social selection on the population dynamics of Huntington's disease. *Ann. Hum. Genet. Lond.* **43**, 413–17.

The emergence, maintenance, and transitions of the earliest evolutionary units

EÖRS SZATHMÁRY

1. INTRODUCTION

When contemplating the origin of life, our *biological* knowledge must be extendable *downward*, and our *physico-chemical* knowledge must be extendable *upward* to reach the no man's land between the realms of molecular chaos and biological order. Only then can a satisfactory explanation of the origin of life emerge.

We are interested in the emergence of the earliest biological individuals, being 'spatiotemporally localized entities that have reasonably sharp beginnings and endings in time' (Hull 1980). The appearance of complex natural systems are conceivable on two different, but not mutually exclusive, grounds: through the operation of non-evolutionary self-organizing processes (such as those described by synergetics: Haken 1983) and evolutionary transformations *sensu stricto* resting on the action of natural selection.

From an evolutionary point of view, organisms consist of two essential parts: replicators and interactors. Whereas replicators are always copied (replicated) through some material process, interactors are not (Hull 1980). In other words, interactors are the vehicles of replicators (Dawkins 1982).

Through the eye of a theoretical biologist interested in organisms, it seems that living beings have two essential subsystems: one ensuring metabolism and homeostasis while consuming material and energy, and another one serving as a programme-controlling device (Gánti 1966, 1971, 1979, 1987; Maynard Smith 1986). The metabolic subsystem is a dissipative one, it assumes a far-from-equilibrium state, maintainable only by energy consumption and entropy production (Bauer 1967; Schrödinger 1967). In contrast, the programme-controlling subsystem, although inactive without energy consumption, is essentially a non-energetic entity. It is the *information* rather than the matter or the energy that matters; widely different programmes can exist at (roughly) the same energetic level.

A unit of evolution must show three basic properties: the capacities for multiplication, heredity, and variation (Maynard Smith 1986, 1987). Multiplication by itself is insufficient; strictly identical self-propagation, although interesting, cannot present the raw material of evolution. Heredity can be given two interpretations: (1) like begets like (important parent–offspring correlation); and (2) the genetic material is copied. The former is a phenomenological, the latter a mechanistic statement. It is generally understood that (2) is the causal factor underlying (1). Even for sexual organisms, resemblance of the child to mother and father is due to the inherited DNA, despite its rearrangements (recombination) and mutations. A highly interesting outcome of theoretical studies related to the origin of life (discussed in section 8) is the idea that an evolutionary unit might exist without *any* copying process (template replication). Variation means that heredity is not perfect (like sometimes begets unlike).

In this chapter I shall primarily concentrate on those aspects of the earliest molecular and life forms which could have enabled them to act as units of evolution. In other words, the origin, maintenance, and evolution of replicators rather than vehicles will be exposed. The surprising results on non-replicative heredity, however, will force me to deal with problems of metabolic networks as far as there can be, in certain cases, non-disentanglable connections between the two basic aspects: metabolism/homeostasis and programme-control.

We are interested here in evolutionary processes leading ultimately to the first soft, programme-controlled automata, i.e. certain kinds of fluid chemical machineries having metabolism, spatial separation, and encoded control (Gánti 1979, 1987), and their earliest development. However, some interesting connections between this—perhaps somewhat esoteric—area and 'conventional' evolutionary biology do exist. I will mention some and elaborate on these if applicable.

The general interest of this area is two-fold. First, the origin of life is one of the major unsolved problems of science (Crick 1982). Secondly, the emergence of increasingly higher-level evolutionary units from lower level ones has received revived interest (Buss 1988).

In Section 2, the nature and modelling of RNA replication is discussed, followed by some interesting features of non-enzymatic template replication and the coexistential consequences of sub-exponential growth (Section 3). The hypercycle concept, a leading idea in the field, is presented in Section 4. An unusual feature of the hypercycle is that, in its simplest form, growth is hyperbolic rather than exponential. The possible implications for the uniqueness of the code and chirality (the handedness of biomolecules) are discussed in Section 5, but I also argue that more realistic models of the hypercycle need not have hyperbolic growth. Section 6 discusses the evolution of the hypercycle, and argues that it can only serve as an effective integrator of information if its components are enclosed within compartments, or protocells. Then I describe an alternative idea, the stochastic corrector mechanism as a means of building novel evolutionary units (Section 7). In Section 8, I show how compartmentalized, reflexively autocatalytic units could behave as units of evolution without complementary template copying. Finally (Section 9), I apply the principles that I have presented to suggest an experimental test-system for systematic production of enzymatic RNAs (ribozymes).

This chapter partly complements that of Bulmer (1988) in the preceding *OSEB* volume. A related, compact review is presented by Szathmáry (1989).

2. THE IMPASSE OF REPLICATING RNAS: EIGEN'S PARADOX

The most simple evolutionary units presently encountered are RNAs replicated in the test-tube (Spiegelman 1970). This experimental system is derived from the Qß phage of *Escherichia coli*. Qß is an RNA virus replicating without a DNA intermediate. The multi-subunit replicase enzyme is partly encoded by the phage's genetic information, partly borrowed from the host's set of proteins. It is capable of copying single-stranded RNA if activated nucleotides are present. The initial (or +) strand is used as a template to produce the complementary (or −) strand, then vice versa.

If one starts the experiment with the complete phage RNA, the size of the template is quickly reduced from approximately 4500 to approximately 200 nucleotides. This is understandable because the major part of the original template is not necessary for the replication cycle in the given (artificial) environment. (The situation is analogous to that in which cave-dwelling animals evolve a decreased capacity for vision.) As the replicase is provided, the template does not have to code for anything any longer; only the so-called target function matters; i.e. the RNA must be a good substrate of the replicase, meaning that it must be recognized efficiently and replicated quickly by the enzyme.

As RNAs are replicated with mistakes, and the variants can have different reproductive success, RNA molecules *in an appropriate environment* can be regarded as units of evolution. A formal treatment, the quasispecies model (Eigen 1971; Eigen and Schuster 1977) expresses this in an elegant form.

Let the sequences be numbered from $i = 1$ to N. Let the overall replication efficiency of the ith sequence be A_i, its copying fidelity Q_i, and the decomposition (hydrolysis) rate D_i. As all sequences other than i can mutate, through their erroneous copying, to produce type i molecules, we must incorporate the corresponding mutation rates w_{ij} as well. Now Eigen's equation reads:

$$dx_i/dt = (A_iQ_i - D_i)x_i + \sum_{j \neq i}^{n} W_{ij}x_j - x_i\Phi \qquad (1)$$

where x_i is the concentration of the ith sequence and Φ is an outflow term ensuring that:

$$\sum_{i=1}^{N} x_i = 1,$$

(2)

which is a selection constraint (it is known that the total concentration can be taken as unity without loss of generality). We assume that the concentrations of the raw materials are buffered so that they can be incorporated into the rate constants. $W_i = A_i Q_i - D_i$ is called the fitness of molecular type i. For the sake of simplicity, direct copying is assumed instead of the $+/-$ complementary replication.

What can we say about the nature of the emerging molecular distribution? Theoretically, all sequences are connected by positive mutation rates. Assume that there is a single sequence type (the master) with the highest A_i, and that decay rates are negligible. If the copying fidelity (Q_i) is not too low, the master sequence will coexist with a distribution of its mutants. This stable distribution together with the master corresponds to the winning quasi-species, which was given a precise mathematical definition (Eigen and Schuster 1977).

First let us consider, following Maynard Smith (1983), a simplified version of eqn (1) with two sequence types only, the master (subscript 0) and its mutant (subscript 1):

$$x_0' = AQx_0 - x_0(Ax_0 + ax_1)$$

(3a)

$$x_1' = ax_1 + A(1-Q)x_0 - x_1(Ax_0 + ax_1)$$

(3b)

where the prime indicates the time derivative, and it is assumed that there is no back-mutation to the master, and that decay rate constants are zero. Note also the construction of the outflow term, ensuring that $x_0 + x_1$ is constant. We assume $A > a$.

We are interested in the case when both species coexist in selection equilibrium. To see this one can set both equations equal to zero and divide the first one by x_0 and the second one by x_1. As the outflows are equal, we obtain:

$$AQ = a + A(1-Q)\hat{x}_0/\hat{x}_1$$

(4)

from which, after rearrangement and considering eqn (2), we have:

$$\hat{x}_0 = (AQ - a)/A - a)$$

(5)

which is positive if:

$$Q > a/A = 1/\sigma$$

(6)

where σ is by definition the selective superiority of the master. This equation indicates that the overall fidelity of the master must exceed a certain limit.

This result can be worked upon further (Eigen 1971). Assume that the master consists of n nucleotides and the average copying fidelity *per nucleotide* is q. Then we have:

$$Q = q^n \approx e^{-n(1-q)} \tag{7}$$

where the last formula is the first member of the Poisson distribution. Combining the last two equations we obtain:

$$n < \ln\sigma(1-q) \tag{8}$$

Thus the number of nucleotides is confined to values determined by the selective superiority, and by the accuracy per nucleotide. The influence of the latter parameter is more severe as the former appears only in the logarithm.

More elaborate treatments (Swetina and Schuster 1982) show that an error threshold also occurs for the more general system, eqn (1). For error rates above the threshold, the population is no longer localized around an optimum in sequence space, but sequences occur in frequences determined by the mutation rates w_{ij}, and not their selective values, W_i.

The problem is reminiscent of Muller's ratchet (cf. Maynard Smith 1978), in which an asexual population deteriorates if the genome size is too large relative to the number of individuals and the mutation rate. Note, however, that Muller's ratchet owes its effect to sampling error, or stochasticity, whereas the inequality, eqn (8), is strictly deterministic. It should be noted that the Eigen eqn (1) can be regarded as a full model of a haploid asexual population with N alleles (Maynard Smith 1983).

Verbally, the meaning of the threshold is the following. If the molecule is too long, many errors accumulate in a single round of replication. The proportion of correctly replicated master molecules becomes so small that the *net* growth in density will decrease below that of the mutants, the latter owing their superiority to the high mutation rate from the master and the high interconversion rate within the mutant distribution (negligible backflow to the master).

Condition (8) has far-reaching consequences, As Eigen (1971) has pointed out, there are chemical reasons why the single-digit accuracy could not have been greater than 0.99 for the primordial, non-enzymatically replicating RNAs. This makes it very difficult to selectively maintain templates with more than approximately 100 nucleotides (the size category of present-day tRNAs). However, a single such molecule is insufficient to provide the coding capacity for the genome of a protocell. *Several templates are needed with widely different sequences and functions.*

The solution of the problem might seem at first glance to be obvious. The quasispecies can involve all sorts of sequences. The problem is that they will not be represented in sufficient quantities. Imagine that one element of the basic genomic set is the master itself. Close relatives to the master cannot have widely different functions as their sequences are not different enough. On the other hand, mutants far from the master will be represented in negligible concentrations due to their low fitnesses. In fact, the equilibrium concentration can be so small that it means only a fraction of a molecule in a few cubic metres. In this case deterministic description of the dynamics invariably breaks down; a stochastic treatment is necessary (Swetina and Schuster 1982). Beside this technical remark, the major conclusion is that the *non-enzymatically* replicating RNA quasispecies is not a sufficient information integrator.

Enzymes are able to raise the error threshold considerably (Table 1). It can be seen that bacteria owe the maintenance of their large replicated genome to the proof-reading capacity of the DNA replicase machinery. The genome of many eukaryotes is certainly beyond the threshold allowed for the intrinsic accuracy of replication. It has been conjectured (Eigen and Schuster 1977) that it is recombination which enables such large genomes to persist. However, this is an open question calling for clarification. First, it should be noted that there exist asexual organisms in the whole range of genome sizes

Table 1

Selectively maintainable information storage capacity of various genetic systems.
(Reproduced from Eigen and Schuster 1977)

Single digit error rate $(1-q)$	Superiority (σ)	Maximum length in digits (n_{max})	Molecular mechanism examples
5×10^{-2}	2	14	Enzyme-free
	20	60	RNA replication
	200	106	(tRNA $n=80$)
5×10^{-4}	2	1386	RNA replication
	20	5991	via specific
	200	10 597	replicases, Qβ phage, $n=4500$
1×10^{-6}	2	0.7×10^6	DNA replication
	20	3.0×10^6	with proof-
	200	5.3×10^6	reading and mismatch repair *E. coli*, $n=4 \times 10^6$
1×10^{-9}	2	0.7×10^9	DNA replication
	20	3.0×10^9	and
	200	5.3×10^9	recombinational repair, man, $n=3 \times 10^9$

and, secondly, that large genomes may have nucleosceletal rather than coding or regulating function (Cavalier-Smith 1985). It is the *actually coding portion* of the genome that matters in the error threshold problem. I know of no investigation in the literature pursuing the analogue of eqn (8) when recombination is effective. Certainly, the generalization of the Eigen eqn (1) to a recombinative system is by no means straightforward nor easy. Whereas eqn (1) is analytically solvable with the aid of a time-varying transformation (Jones *et al.* 1976), resulting in a linear system, recombination makes the system inherently non-linear due to the mating 'reaction' (two replicators must find each other for recombination to occur).

To summarize, there is a limit on the size of the genome that can be replicated, given approximately by the equation:

$$n < \ln\sigma/(1-q),$$

where n is the number of nucleotides, σ a measure of the selective advantage of the fittest sequence, and $1-q$ the probability of an error per base replication. It follows that non-enzymic replication of RNA could not, on its own, sustain sufficient genetic information for the first evolutionary units.

3. NON-ENZYMATIC REPLICATION AND THE SELECTION CONSEQUENCES OF THE SUBEXPONENTIONAL GROWTH LAW

Although non-enzymatic replication has often been assumed, the experimental demonstration of template replication and recycling has only recently been demonstrated (von Kiedrowski 1986; Zielinski and Orgel 1987).

In this section, I first show that, in a population of non-enzymatically replicating molecules, the growth rate of a particular template is less than exponential—in fact, it follows a square root law. It follows that, if there are two competing types, the likely outcome is coexistence, rather than the selective elimination of one of them. The reason is that the template will readily form a duplex with its complement, and is then temporarily unavailable for replication. Thus, each template interferes with its own replication (by duplex formation) more than it interferes with the replication of its competitor: ecologists will recognize that this favours the coexistence of competitors. The resulting molecular variability has interesting consequences for early evolution.

The phenomenological rate equation describing the growth of total template chain concentration (x) is:

$$x' = k\sqrt{x} \tag{9}$$

To understand the growth kinetics, suppose that the concentration of the free single-stranded template be y, that of the double-stranded form z. Replication is possible only if the duplex opens itself (Fig. 1). However, the association–dissociation balance is strongly shifted to the duplex form, i.e. the kinetic rate constants obey the:

$$a \gg b \tag{10}$$

relation. Replication is the slowest process, so that $b > k$. In this case the association–dissociation process assumes a quasi-equilibrium state, where the reaction rates in the two directions are equal:

$$ay^2 = bz \tag{11}$$

The total concentration is $x = y + z$. From eqn (11) we have:

$$y = \sqrt{(bz/a)} = \mathrm{f}\sqrt{z} \tag{12}$$

where f is a constant. As eqn (10) is valid, y/z is very small, or, in other words, $x \approx z$. Putting this into eqn (12) we arrive at the conclusion that the concentration of the single-stranded form approximately equals the square-root of the total concentration. Since only this form is active in replication, eqn (9) must be approximately valid as well.

Referring to Eigen and Schuster (1977), von Kiedrowski (1986) assumes that the dynamic eqn (9) implies Darwininan behaviour ('survival of the fittest') when different sequences are allowed to compete in the same environment. This is a mistake, however (Szathmáry and Gladkih 1989) Consider first the general growth equation:

$$x' = kx^p \tag{13}$$

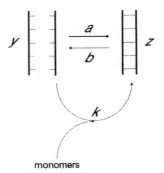

Fig. 1. The relation of single- and double strands in non-enzymatic template replication. Rate constants and concentrations corresponds to those in the text.

$p=1$ corresponds to exponential (Malthusian) growth. If $p>1$, we have hyperbolic growth, of which I will write later. $0<p<1$ corresponds to subexponential but supra-linear growth.

Let us now construct a selective system using eqn (13). Imagine an 'evolutionary reactor' in which the concentration of raw materials is buffered and the total concentration of the templates is kept constant by an appropriate non-specific outflow (dilution). This is the constraint of constant total concentration (Eigen 1971). I present the equations for two species only; generalization to more types is straightforward.

$$x_1' = k_1 x_1^p - x_1(k_1 x_1^p + k_2 x_2^p) \tag{14a}$$

$$x_2' = k_2 x_2^p - x_2(k_1 x_1^p + k_2 x_2^p) \tag{14b}$$

This system ensures that if initially $x_1 + x_2 = 1$, it will not change (Fig. 2). It is also known that this assumption does not affect the generality of the conclusions on selection (Eigen and Schuster 1977).

First let us see if coexistence ($\hat{x}_1, \hat{x}_2 > 0$) is possible. To this end, set the time derivatives equal to zero and divide eqns (14a) and (14b) by x_1 and x_2, respectively. Noting that the outflow terms are equal, we obtain:

$$k_1/k_2 = \hat{x}_1^{(1-p)}/\hat{x}_2^{(1-p)} \tag{15}$$

It is apparent that if $p=1$, eqn (15) is only possible if $k_1 = k_2$, echoing Gause's (1934) principle: complete competitors cannot coexist (when growth is Malthusian). In the subexponential case there exists an internal equilibrium point; the question is, is it stable or not? It turns out that it is; I show this by showing that both species can invade when rare. We write eqn (14a) in a different form:

$$x_1' = x_1^p(k_2 - x_1^{(1-p)})(k_1 x_1^p + k_2 x_2^p) \tag{16}$$

Growth is possible if the first bracketed expression is positive, which always

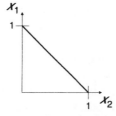

Fig. 2. The meaning of the constraint of constant total concentration (c) in the case of two variables. The system is confined to the straight line when $c=1$.

holds if x_1 is sufficiently small. By symmetry, the same applies to eqn (14b). *Subexponential growth implies coexistence.*

One must not forget that eqn (9) is a result of an approximation. A complete dynamics should include separate equations for at least the single- and double-stranded forms. A detailed model reveals that coexistence is still possible if the total concentration is not too small (otherwise the single strands do not find each other and cannot give rise to the duplex).

The potential importance of these theoretical results is as follows. First, it makes the prediction that experimental demonstration of coexistence of different non-enzymatically replicating sequences should be possible. Secondly, this mechanism could have provided a means to present the primordial cells with a sufficient variety of oligoribonucleotide modules of which the larger molecules, with diverse sequences, could have been fabricated.

There is suggestive evidence and speculation that primordial t- and rRNAs, and even mRNAs were made of shorter oligoribonucleotide modules. I will discuss these items in turn.

Ujhelyi (1983) has shown by sequence comparisons that tRNAs and 5S rRNA share striking similarities and possibly a common ancestry. By a more involved analysis Bloch *et al.* (1985) have shown that a repeated and later modified 9 ± 1 nucleotide module possibly underlies the discovered homologies. tRNAs and rRNAs could have been fabricated from these modules by self-priming and recombination (Fig. 3).

According to Ohno (1987), base oligomers could have played a crucial role in the origin of the earliest protein-coding sequences as well. There are three arguments in favour of this hypothesis. First, short sequences with internal duplications could efficiently elongate themselves by shifting and re-annealing (Fig. 4). Secondly, polypeptides encoded by these repetitive sequences would have been periodic themselves, folding into α-helical or β-pleated-sheet secondary structures more easily. Thirdly, if the length of the repeated unit is not an integer multiple of 3, the coding sequence could be immune to random base substitutions, insertions, and deletions. If one reading frame on the RNA is open, so must be the other two (Fig. 4).

Although it is known that the possibly very ancient sugar-metabolizing enzymes are almost completely built of α-and β-structures, searching for internal periodicities in the messengers, as has been done for t- and rRNAs, would be welcome.

Ohno's mechanism is not the only one to solve the problem of reading frame determination and premature cessation of translation in a primitive protein sythesizing machinery. Crick *et al.* (1976) and Eigen and Schuster (1978b) have pointed out that an RRY or an RNY codon pattern (R = purine, Y = pyrimidine, N = any base), respectively, would ensure automatic preservation of the reading frame without sophisticated ribosomal reading (see also Bulmer 1988). A 5'-RNYRNYRNY-3' messenger is necessarily read in the correct frame by adaptors having 5'-RNY-3' anticodons.

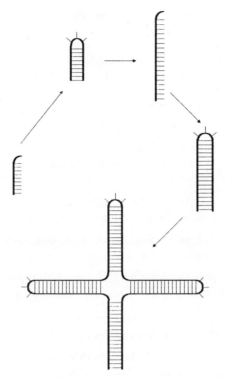

Fig. 3. The formation of tRNA-like structures by self-priming and self-templating of short oligoribonucleotides (Bloch *et al.* 1985).

What makes Ohno's suggestion appealing is that it does not restrict the extent of the code vocabulary. All this, of course, is connected to the problem of the code's origin, very briefly discussed in Section 9.

The question naturally arises how the results on non-enzymatic replication kinetics relate to those on enzymatic replication. Epstein (1979) has shown that *replication* with *parallel* non-replicative pairing can lead to template coexistence; the number of coexisting species increasing with total concentration. Biebricher *et al.* (1983, 1985) have considered competition of *enzymatically* replicated RNA species, based on elaborate experiments. They also treated pair formation as an *irreversible* process, and, in addition to homologous pairs, they took hybridization between two *different* sequences into account. They have found that in the *linear* growth phase, when the Qβ replicase becomes saturated due to high template concentration, homologous pairing begins to dominate the dynamics and leads to coexistence.

The crucial difference between the non-enzymatic and the enzymatic models is as follows. In the latter, the duplex does not dissociate ($b = 0$ in Fig. 1), and replication gives rise to single-stranded rather than double-stranded forms, due to the 'skilful' activity of the enzyme. A protobiological system is

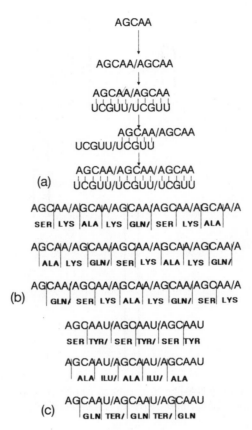

(a)

(b)

(c)

Fig. 4. The growth and coding capacities of oligomeric repeats (Ohno 1987). (a) The pentamer is elongated by unequal pairing and primed replication. (b) The elongated RNA codes for a polypeptide chain with the pentapeptidic repeat Ser–Lys–Ala–Lys–Gln in *all three* reading frames. This is due to the fact that the initial oligomer is a pentanucleotide. (c) The one nucleotide longer hexanucleotide repeat can code only for dipeptidic repeats, or may be aborted by terminator codons (ter). Thus, ancient RNAs are preferably made of oligoribonucleotide modules whose sizes are not integer multiples of three.

unlikely to use this type of enzymatic kinetics as an information integrator because it is very wasteful: it requires very high total template concentration, and replication is a process completely separate from pair formation. Despite the same 'ecological' background (stronger self-limitation than cross-limitation), the mechanistic details differ, requiring other ranges of parameter values for coexistence.

Thus we have found that the coexistence of templates resting on the subexponential growth law could have been useful in the preservation of a panoply of RNA (or RNA analogue) modules, but is unlikely to have contributed to the integration of longer sequences.

4. THE HYPERCYCLE: A PRINCIPLE OF NATURAL SELF-ORGANIZATION

In order to overcome the difficulties associated with the error threshold, Eigen (1971) has introduced the concept and model of a hypercycle (Fig. 5). Each of its members is an autocatalytic template, but in addition to this, each member provides a (hetero)catalytic aid in the replication of the cycle's next member. The catalytic aid may be direct or indirect, the latter accomplished by encoded protein replicases. For the hypercycle with translation (the 'realistic hypercycle') it is tacitly assumed that the RNA members have triple functionality: they are messengers, templates, and adaptors at the same time (Eigen and Schuster 1978*b*).

The hypercycle has been supposed to meet the criteria for *the integration of genetic information* dispersed in otherwise competitive replicators (Eigen and Schuster 1977, 1978*a*). For this to be true, the following conditions must be satisfied:

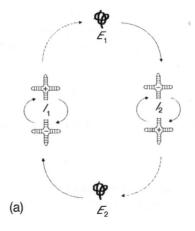

(a)

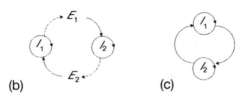

(b)   (c)

Fig. 5. The hypercycle. (a) Detailed mechanism: I_i are information carriers, E_i encoded replicase enzymes. Single arrows indicate catalytic action, dashed arrows indicate translation. (b) The same as (a) in a more abstract form. (c) This simple version may express either a further simplification of (a) and (b), or may indicate a hypercycle without translation i.e. based on ribozymes I_i which can act directly as specific replicases.

1. In the new system, members must still be capable of successful competition against their error copies.
2. Competition between the members to be integrated must be suspended (the possibility of competitive exclusion is excluded).
3. The whole functional system must favourably compete with less efficient systems or single replicators.

I now show that criterion 2 is fulfilled by the hypercycle. Again, I restrict myself to two members (see Eigen and Schuster 1978a for details). Consider the system below:

$$x_1' = k_1 x_1 x_2 - x_1(k_1 + k_2)x_1 x_2 \tag{17a}$$

$$x_2' = k_2 x_2 x_1 - x_2(k_1 + k_2)x_1 x_2 \tag{17b}$$

where x_1 is the concentration of the ith member. Again, we imagine the hypercycle to be dissolved inside an evolution reactor in which the total concentration is constant and unity (again without loss of generality).

It is easy to see that coexistence is always possible, irrespective of the value of k_1/k_2. Set eqns (17a) and (17b) equal to zero and divide the first equation by x_1 and the second by x_2. Then we have:

$$k_1/k_2 = \hat{x}_1/\hat{x}_2 \tag{18}$$

This equilibrium point is stable. Rewriting eqn (17a) we obtain

$$x_1' = x_1 x_2(k_1 - (k_1 + k_2)x_1) \tag{19}$$

which is positive if x_1 is sufficiently small; thus member 1 can invade when rare. By symmetry the same is true for member 2. Hence we see that the members of the hypercycle are co-operative.

Equation (17) shows that no member can replicate when the other is absent. Yet this is only a necessary but not sufficient condition for co-operation (Eigen and Schuster 1978b). To see this consider the following system:

$$x_1' = k_1 x_1 f(x_1, x_2) - x_1(k_1 x_1 + k_2 x_2)f(x_1, x_2) \tag{20a}$$

$$x_2' = k_2 x_2 f(x_1, x_2) - x_2(k_1 x_2 + k_2 x_2)f(x_1, x_2) \tag{20b}$$

where $f(x_1, x_2)$ is a function assuming zero value when either concentration is zero. This system can be considered a crude model of the situation in which both templates are necessary to run a common metabolism, but the benefit gained through that is not specifically channelled back to either of them (Fig. 6).

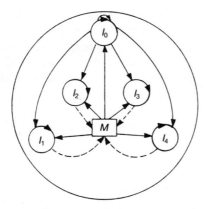

Fig. 6. Scheme of a system with metabolic function. Template I_0 is coding for a replicase activity, the information of templates I_1-I_4 is necessary to run the metabolism M, which in turn is necessary for template replication. Single arrows indicate participation in replication, dashed arrows indicate contributions to metabolism. Despite all these connections, the fastest replicating template outcompetes the others. A similar system with a similar fate is shown by Eigen and Schuster (1978b). (Reproduced from Szathmáry and Demeter, 1986.)

Note how the above system resembles eqn (14) with $p=1$. It is apparent that eqn (20) can be regarded as a competition between Malthusian replicators with time-varying rate constants in which the ratio of the constants does not alter. It is easy to check by the methods used here that in fact there is no coexistence in eqn (20) unles $k_1 = k_2$ (the trivial case). Hence, the stable coexistence of two replicators in such non-hypercyclic systems is not guaranteed, even if each is essential to the replication of the other.

To avoid confusion, I must emphasise (Szathmáry 1988) that a simple chemical cycle (such as the Szent-Györgyi–Krebs cycle) does not qualify as a hypercycle only because each step in it is catalysed by an appropriate recycling enzyme. Such systems should be called supercycles (Gánti 1984) or secondary cycles (Szathmáry 1988). The crucial difference between these and hypercycles is the auto- and heterocatalytic nature of each member in the latter system.

5. HYPERCYCLIC COUPLING AND HYPERBOLIC GROWTH

One interesting feature of the *simple* hypercyclic growth equation is that it is hyperbolic rather then exponential (Eigen 1971; Eigen and Schuster 1978a). It means that the total concentration $x = x_1 + x_2$ obeys a growth law eqn (13) with $p > 1$. The selection consequence is 'once-for-ever' (Eigen 1971) or, in a more appropriate formulation, 'survival of the common' (Michod 1983, 1984). This is due to the fact that a system growing hyperbolically reaches infinite concentration in *finite* time, whereas exponential growth reaches infinite concentration in infinite time only (it is assumed that the reservoir of raw materials is always infinite). To show this let us integrate eqn (13) with $p > 1$. Elementary calculus gives:

$$x(t) = [x(0)^{(1-p)} - kt]^{1/(1-p)} \qquad (21)$$

or, with $p = 2$:

$$x(t) = 1/[1/x(0) - kt] \qquad (22)$$

which apparently approaches infinity as $t \rightarrow 1/[x(0)k]$. This critical time interval is smaller for larger k. Note that explosion time depends not only on 'fitness', but on initial concentration as well, leading to survival of the common in a competitive situation. To see this directly consider competition of two hyperbolically growing species:

$$x_1' = k_1 x_1^2 - x_1(k_1 x_1^2 + k_2 x_2^2) \qquad (23a)$$

$$x_2' = k_2 x_2^2 - x_2(k_1 x_1^2 + k_2 x_2^2) \qquad (23b)$$

Our standard method quickly shows that coexistence is possible if $k_1\hat{x}_1 = k_2\hat{x}_2$. This point is unstable, however. Re-writing eqn (23a) we have:

$$x_1' = x_1[k_1 x_1 - (k_1 x_1^2 + k_2 x_2^2)] \qquad (24)$$

showing that if $x_1 \rightarrow 0$ (and, consequently, $x_2 \rightarrow 1$) the concentration of species 1 is bound to decrease, meaning that it *cannot* invade when rare. By symmetry, the same holds for species 2 as well. The stationary points $(0, 1)$ and $(1, 0)$ are stable locally. Thus, the species with the largest $k_i x_i$ value wins the competition. Eigen and Schuster (1978a, p.41) conclude: '*For hypercycles, selective advantages are always functions of population numbers, due to the inherently nonlinear properties of hypercycles . . .*' (their italics).

One ought to realize that this is not a problem of marginal importance. As Küppers (1983, p. 209) writes: 'In particular, a catalytic hypercycle, once selected, will not tolerate (in homogeneous solution) the nucleation of independent competitors. The genetic code and the chirality of systems which have emerged from a single hypercycle are therefore universal.'

As it has been shown (Szathmáry 1984; Szathmáry *et al.* 1988) hypercycle theory lacks internal consistency in this regard. First of all one should not forget that the hypercycle is a catalytically closed entity. This means that the members could, in principle, replicate by themselves as pure autocatalysts, following a first-order growth law. Indeed in the same publication, Eigen and Schuster (1978a, p. 15) remark that 'at the low concentration limit ...' a hypercycle, the members of which undergo simple replication as well, becomes '... identical with the system of exponentially growing, independent competitors.' It is very easy to see why this is so. Consider the equation:

$$x_i' = k_i x_i + k_i x_j x_i, \ i = 1 \text{ to } N \tag{25}$$

where j is always the preceding member of the hypercycle relative to i. It is apparent that as the concentrations approach zero, the second term becomes negligible relative to the first one, thus leading to exponential competition.

Yet the story does not end here. Also in the same publication Eigen and Schuster (1978a) elaborate more detailed kinetic models of the hypercycle. Specifically, the association–dissociation process of the replicase–template complex is taken into account with the conclusion: 'Thus, under saturation conditions, i.e. at high concentration of the constituents, the hypercycle loses the behaviour typical of nonlinear growth rates' (p. 33).

It has been suspected that the unification of these two 'exceptions' would be devastating for the hyperbolic growth 'law' of hypercycles (Szathmáry 1984). Szathmáry et al. (1988) have therefore attempted an exact analysis, complementing King's (1981) earlier approach. The dynamical behaviour depends critically on whether the individual replicators depend absolutely on the other members of the cycle—'obligate' hypercycles—or whether they are capable of independent replication—'facultative' hypercycles. In the case of obligate hypercycles survival of the initially common form (masking of intrinsic fitness by the growth law: Michod 1984) remains valid, in the sense that a very low concentration hypercycle cannot invade. However, for facultative hypercycles, the masking of intrinsic fitness and the victory of the initially common is not a necessary feature, although it can hold for some parameter values. The dynamics are complex: it is even possible that two facultative hypercycles can coexist at a locally stable equilibrium, even if one system is stable against invasion by the other.

Eigen and Schuster (1982) assume at the beginning of their scenario that a population of autocatalytic templates exists and it is within this population that hypercyclic coupling develops. In this case initial heterocatalytic (replicase) activity would have been very low, giving rise to facultative hypercycles and, therefore, the possibility of coexistence. As we have seen, survival of the common is not a general rule for hypercycles as such. It follows that ancestral universality of the genetic code or of chirality is not readily explained by assumed hypercyclic organization (Szathmáry et al. 1988). An even more serious objection will be dealt with in the next section.

6. THE GENETIC INSTABILITY OF 'NAKED' HYPERCYCLES

The hypercycle can fall victim to at least three catastrophes (Fig. 7): selfish RNAs (efficient parasites), short-circuit mutants, and disappearance of members due to fluctuations (Bresch *et al.* 1980; Niesert *et al.* 1981; Dyson 1985). The first two are more likely to occur at high population density (mutants arising more frequently), while the third menace increases with decreasing population numbers; hence a system must sail between Scylla (high density) and Charybdis (low density) like resourceful Odysseus (Niesert *et al.* 1981). The point is that naked (non-compartmentalized) hypercycles just cannot avoid the perilous reefs.

A crucial difficulty concerns the kinds of mutation affecting hypercyclic function that can be established by selection. Maynard Smith (1979) distinguished between (i) mutants altering a template so that it is a better (or worse) replicator, while encoding an unaltered replicase, and (ii) mutants leaving the template unaltered as a replicator, but encoding a replicase with altered properties. This corresponds to the distinction made by Eigen *et al.* (1981) between 'phenotypic'—type (i)—and 'genotypic'—type (ii)—mutations. The importance of the distinction is as follows (Fig. 8). Type (i) (phenotypic) mutants are affected by selection, since mutants that are better replicators will increase in frequency, whereas type (ii) (genotypic) mutants are not selected for or against. This distinction will be familiar to evolutionary biologists in another context: type (i) mutants are 'selfish' mutants, altering the individual fitness of the template, whereas type (ii) mutants affect only the reproduction of others. In an unstructured population, type (ii) mutants are selectively neutral. The spread of type (ii) mutants therefore requires that the replicators be enclosed within compartments.

Eigen *et al.* (1980), replying partially to criticism, point out that in the initial phase of hypercycle evolution at least the target efficiency—a phenotypic trait—is optimized, and immediately after this the hypercycle has to escape into a compartment. In conclusion, it has been said that compartments may assist—but do not replace—hypercycles.

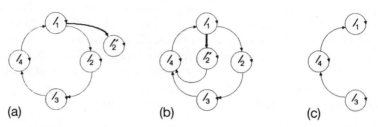

Fig. 7. Hypercyclic catastrophes. (a) A complete parasite. (b) Short-circuit. (c) Fluctuative loss of one member. Bold arrows indicate increased catalytic strengths.

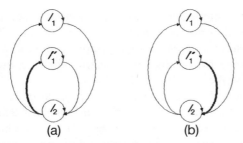

Fig. 8. Mutants in the hypercycle. (a) A phenotypic mutant selects itself through selective feedback. (b) A gentotypic mutant cannot be selected for, as specific feedback is lacking. For the sake of simplicity, the principles are illustrated on the non-translational hypercycle. Bold arrows indicate increased coupling strengths.

This view has been challenged by Szathmáry and Demeter (1987). First it can be shown that optimization of phenotypic traits is less than enough. Consider two mutation scenarios of a simplified (so-called elementary, non-translational) hypercycle (Fig. 9). In the first scenario the historically second mutant is a descendant of the first mutant. As we see, target efficiency increases, although one among the two catalytic activites is allowed to decrease in strength arbitrarily, this process could end with the evolution of a

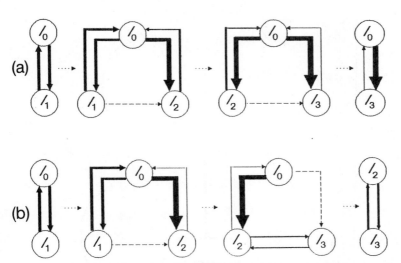

Fig. 9. Two possible scenarios of hypercyclic mutations. (a) The case of ever-increasing target efficiency (subsequent selection of selfish mutants). Note that the other coupling has become weaker. (b) Here the two mutants arise in such a way that finally both catalytic activities decrease in strength, due to *positive* selection for target efficiency in each step. Broken arrows indicate mutations, single arrows indicate catalytic couplings; boldness is proportional to strength.

template that was completely parasitic. In the second scenario, the historically second mutant is *not* an offspring of the first one. This leads to the discomforting result that *both* catalytic couplings are allowed to decrease in strength. One has to realize that this scenario works through the optimization of the phenotypic trait (target function).

The conclusion is that the naked hypercycle is an inefficient information integrator for it violates the third criterion of Eigen and Schuster (1978*a*; see above). The essential reason for this was realized by Gánti (1978, p. 67): '... one should note that the Eigen theory does not provide details on the spatial separation, or individuality, of hypercycles, which leads to a cross-flow of information among the different species, confusing the evolutionary process.' Naked hypercycles are not individuals, hence they cannot be units of evolution.

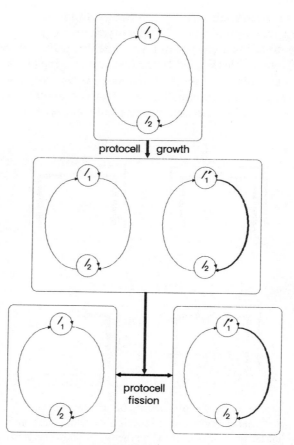

Fig. 10. Compartmentalized (clothed) hypercycles as Darwinian units of evolution. The genotypically favourable mutant can be segregated from the wild type, so that it can be selected for.

Of course, clothed hypercycles (with compartmentation) are a sufficient solution to the problem. Imagine that the hypercycle is bounded by a membrane, and the fission rate of the whole compartment depends on the internal composition of the vesicle. It is apparent that now we have a unit of selection (Fig. 10), for 'bad' hypercycles can be distinguished from 'good' ones.

If it is true that hypercycles must go with compartments, chirality and the genetic code cannot be given a survival-of-the-common-type explanation; clothed hypercycles are simple Darwinian units of evolution, with survival of the fittest.

7. THE STOCHASTIC CORRECTOR: ANOTHER PRINCIPLE OF NATURAL SELF-ORGANIZATION

As we have seen, naked hypercycles cannot be efficient information integrators or—which is the same—units of evolution. Compartmentalized hypercycles are satisfactory. Given the possibility of compartmentation, the question naturally arises: hypercycles need compartments, but do compartments need hypercycles? In other words, do protocells have to have hypercycles inside (Szathmáry and Demeter 1987)?

To my knowledge, the first—negative—answer to this question comes from King (1982). In his model he assumes that the presence of several different RNA species is necessary for protocell growth and division. He allows for competition between RNAs, i.e. replication rates are not equal. It is assumed that one copy of each RNA is optimal (for its information can be amplified by translation), and the cellular growth law penalizes unbalanced cells. Cells are assumed to undergo fission once they have reached double size. Each molecule inside is binomially distributed between the two daughter compartments. Despite internal competition, an asymptotic population (a quasispecies of protocells) emerges. As we shall see, this result is due to the *stochasticity* in the distribution of molecules between daughter protocells.

Following an initial idea (Szathmáry 1984), the stochastic corrector model was built as a sufficient alternative to hypercycles (Szathmáry 1986a; Szathmáry and Demeter 1987). Imagine a protocell (protobiological compartment) in which metabolism takes place. We assume, for the sake of simplicity, that two different RNA species are necessary and sufficient to catalyse it directly (RNA enzymes = ribozymes), or indirectly (mRNA and encoded proteins). Without a functioning metabolism RNAs cannot replicate. If they replicate, they do it at unequal rates. Thus consider the following pair of equations (Szathmáry and Demeter 1987):

$$x' = ax(xy)^{\frac{1}{2}} - dx - x(x+y)/k \qquad (26a)$$

$$y' = bx(xy)^{\frac{1}{4}} - dy - y(x+y)/k \tag{26b}$$

where it holds that:

$$a > b \tag{26c}$$

and d is a non-specific decomposition rate, and k is a constant influencing the highest (non-equilibrium) 'carrying capacity' of the system. Comparison of this system with eqn (20) reveals that, although it is true again that the templates cannot replicate alone, the system is non-hypercyclic since there is no locally stable fixed point where the components would coexist. (The power $\frac{1}{4}$ reduces the non-linearity of the system for computational reasons.)

For the above system it is true not only that the templates need each other's presence for replication, but that they *become extinct due to competition*. This result is surprising, but logical. Starting from sufficiently low concentration, first both concentrations grow but, due to (33c), x grows faster. This increases competition between the two templates, which makes y decrease sooner or later. But, because y is necessary for the growth of x, the latter is bound to decrease ultimately as well. The final outcome is that both species disappear in the limit (this can be shown analytically). This fate makes sense from a mechanistic point of view: if we assume that both templates are needed for metabolism, competitive exclusion should lead to overall death.

Now let us put this seemingly hopeless system into compartments. We assume that the fission rate of compartments depends on their internal template composition, with some composition being optimal. In the context of microspheres, application of eqn (26) is no longer justified. Because templates are present in relatively low numbers, concentrations should be replaced by numbers of molecules, and the deterministic dynamic by a stochastic counterpart. The technique adopted is that of master equations. Let the numbers for the two template species be n and m, respectively. The master equation describes the time evolution of the probabilities $p(n,m)$.

In the stochastic corrector model, compartments are distinguished by their initial template composition. From the expected (deterministic) dynamics of eqn (26), we calculate the generation time for each compartment type (non-viable compartments have an 'infinite' generation time) to replace a continuous distribution of generation times for each compartment type. A compartment is assumed to divide if a metabolic substrate s reaches a critical concentration s_{max} starting from $s_{max}/2$. The dynamics of s (given explicitly in the reference) is autocatalytic (production of new metabolites needs the presence of metabolites), and x as well as y also show up in a catalytic manner.

The probability distribution at the point of compartment fission can be calculated for each compartment type. Templates are assumed to go into offspring compartments randomly, obeying a hypergeometric distribution (sampling without replacement). Some of the offspring compartments will be identical to the ancestor, but some of them will resemble other compartment types in their template composition. These compartments are treated as mutants (Fig. 11).

Everything is now ready for constructing an Eigen eqn (1) *at the compartment level*. Amplification factor A_i for the ith type compartment can be calculated from the obvious relation $A_i = 1n(2)/\tau_i$, where τ_i is the compartment"s generation time. The factors Q_i and w_{ij} can be calculated from the stochastic distribution.

We have arrived at the conclusion that, despite internal competition, selection at the compartment level leads to a stable protocellular quasi-

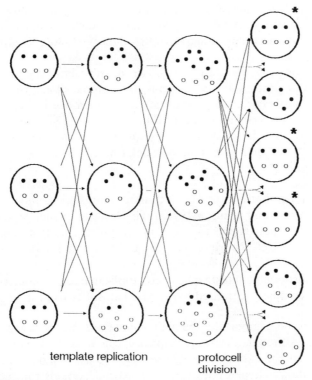

template replication   protocell
                            division

Fig. 11. Schematic representation of the stochastic corrector mechanism. Solid and open circles indicate two different templates, both being necessary for compartment growth. Here the optimal initial composition is three copies of each kind. Replication as well as template allocation in cell division are stochastic processes. Despite the tendency that the solid templates replicate faster, due to stochastic effects the best compartment type (marked by asterisks) recurs. Finally, a self-reproducing compartment distribution (quasispecies) emerges. Reproduced from Szathmáry (1989).

species. This result is due to the fact that all compartment types are dynamically coupled to each other via non-zero mutation rates. Mutations in turn arise from two sources of stochasticity: non-deterministic replication ('demographic stochasticity' in the context of population biology) and chance allocation into offspring compartments ('stochasticity in spatial sampling'). This stochasticity generates variation, on which (at the level of compartments) natural selection acts.

The name stochastic corrector derives from the fact that the deterministic fate of the molecular (and cellular) population is hopeless, but stochasticity corrects this malign tendency by making the best compartment types recur.

At this stage we have two levels of evolution: one is the template level and the other one is the protocell (compartment) level. It can be shown that in the context of the stochastic corrector, selfish (non-catalytic) templates are selected against and better enzymatic functions are selected for. Thus we have a higher-level unit of evolution, with the ability to utilize favourable 'genotypic' mutations. All this suggests that the stochastic corrector model is a sufficient information integrator (Szathmáry 1986b; Szathmáry and Demeter 1987).

Thus it seems that we are left with two possible information integrating mechanisms in protobiological systems: the compartmentalized hypercycle and the stochastic corrector. The question is: which one is the best competitor (F. Hopf personal communication)?

Although a definite answer to this question cannot be given without modelling, my personal bias is on the side of the stochastic corrector, for the following reasons:

1. It is known that hypercycles with more than four members spiral into limit cycles (Eigen and Schuster 1978a). This leads to an increased chance for the fluctuation catastrophe to occur inside protocells, if a stochastic approach is applied (cf. Rodriguez-Vargas and Schuster 1984).

2. The short-circuit catastrophe is not applicable to the stochastic corrector, but the compartmentalized hypercycle can suffer from it.

3. Perhaps most important, the maintenance of hypercycle organization is very costly. Although the same sequence can have two widely different functions in special cases, generally one ought to assume that members of a compartmentalized hypercycle must be bi-cistronic: one cistron coding for an enzymatic function in metabolism, the other one coding for a specific replicase. This burdens the population with two loads: the material one, for information carriers must be about twice the size; and the elevated mutation load, due to the increased length of the templates. To see the last effect quantitatively, note that the overall replication fidelity of a template with n nucleotides is q^n, whereas that of a double-sized template is q^{2n}, assuming

unaltered replicase activity. The ratio of fidelities is q^n. Taking, for example, $q = 0.99$ and $n = 100$, this is 0.36.

It should be noted that the final outcome of selection is neither the hypercycle nor the stochastic corrector. With the improvement of replicase function, and the appearance of DNA, genes can be linked into a single chromosome, in which synchronous replication is guaranteed (Eigen and Schuster 1978b).

Yet another intermediate stage of evolution, somewhat analogous to the technique adopted in present-day eukaryotic S-phase DNA replication, can be conjectured. Eukaryotic chromosomes consist of several replicons, and there is evidence that some bookkeeping mechanism ensures that each one is replicated only once per cell cycle (Alberts et al. 1983). Contrary to speculation (Eigen 1986), it is rather doubtful that this system is hypercyclic, since that would require thousands of replicon-specific, replicon-encoded replicases. Experiments suggest that either a cis-acting negative (Roberts and Weintraub 1988), or a cis-acting positive (Blow et al. 1987) factor may be responsible for bookkeeping.

An intermediate stage in protobiological evolution, where separate RNA genes are still present, but replicated in synchrony by the application of an RNA-bookkeeper, seems to me plausible. Models on this stage are desirable, perhaps following the lines of the genomic tag model by Weiner and Maizels (1987), in which removable tRNA-like tags, together with ancient CCA telomeres, serve as signals for genomic RNA replication. Synchronous replication is a general goal to be achieved during the appearance of novel evolutionary units (Buss 1988), for more faithfully replicated units are at an advantage until copying fidelity becomes reasonably high (cf. the quasi-species model).

On the other hand, the stochastic corrector model has been applied to another megaevolutionary transition, that of the emergence of eukaryotic cells with symbiotically acquired organelles (Szathmáry 1986a). It is well known that autonomously reproducing cell organelles are frequently not in synchrony with nuclear division (Margulis 1981). This creates special problems when two different cell organelles are to be maintained in coexistence within the host cell. It might well be true that both are necessary for the cell (and for each other), but this requirement does not automatically ensure that they do not compete against each other. In fact, the most likely situation is just the opposite: before the evolutionary origin of synchronized reproduction, ancient chloroplasts and mitochondria are likely to have competed with each other in early photosythetic eukaryotic cells. It might even be true that all eukaryotes descend from such cells (Cavalier-Smith 1987). The stochastic correction principle saves the situation: stochasticity generates variation in organelle number, which can be selected upon.

Finally, attention should be drawn to the fact that in the classical body of population genetics, the stochastic correction principle is a case of group selection (cf. Wilson 1983), but in this case *sensu stricto*: 'the entities with the properties of multiplication, heredity, and variation, whose evolution is being described, are in fact the groups, and not individuals' (Maynard Smith 1983, p. 319). Population structure is provided here by protocellular (or, in the other case, eukaryotic cellular) compartmentation.

8. HEREDITY WITHOUT COPYING: REFLEXIVELY AUTOCATALYTIC NETWORKS

Due to the increased understanding of molecular genetics, units of evolution are regarded as based on the template-copying of nucleic acids. It is the goal of this section to call attention to cases for which this does not seem to be true.

Gánti (1979) has imagined the following system. Consider a compartment with metabolism and genes (the so-called chemoton model). The basis of the metabolism is essentially an autocatalytic cycle of intermediates (Fig. 12). Imagine that there are *two* possible cycles; each is able to produce the necessary raw materials for the membrane and the genetic subsystems, but their intermediates are not interconvertible (Fig. 13). If all the intermediates of a cycle are absent, the corresponding enzymes cannot do anything alone.

Now assume that the reactions in cycle A are less sensitive to cold than those of cycle B. If we expose the population to cold treatment, the intermediates of A will take over within the chemotons, and B intermediates become diluted out until all the molecules are lost. The population has *acquired cold tolerance*. As Gánti (1979) remarks, this type of change is not evolution, because it leads to simplification rather than increased complexity. The genes are units of evolution, whereas the described cycles are, at most, units of selection: they multiply, but they do not have variation (cf. Maynard Smith 1987), as the cycles cannot—by definition—'mutate' into each other. Nevertheless, this example reinforces the 'double' character of life: it is not only based on genes (regulatory devices), but on dissipative biochemical networks (regulated devices) as well (Gánti 1975, 1979, 1987; Dyson 1985; Maynard Smith 1986).

Eigen (1971) has considered the possibility of an autocatalytic *protein network* without invoking template replication (Fig. 14). Such a network is possible only if all enzymes are multifunctional, since without complementary instruction, each enzyme needs several others for its production from shorter oligopeptides.

It should immediately be noted that we do not know of such an existing network. There are examples where small oligopeptides are assembled from even shorter ones (e.g. the antibiotic Gramicidin-S) by a multi-enzyme complex. Another interesting case is the 'plastein' reaction, in which the

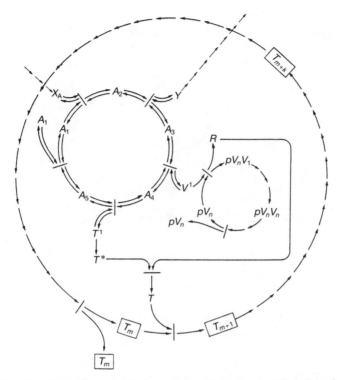

Fig. 12. The minimum chemical network of the chemoton, a well-designed protocell model. The metabolic subsystem, with intermediates A_i, is an autocatalytic chemical cycle consuming X as nutrient and producing Y as waste material. The pV_n macromolecule undergoes template polycondensation (replication), R the condensation by-product is necessary to turn T^1 into T, the membranogenic molecule. The symbol T_m represents a bilayer membrane composed of m pieces of T molecules. It can be shown that this system is able to grow and divide spontaneously. Reproduced from Gánti (1984).

digestive enzyme pepsin, in the absence of high-energy compounds, catalyses the transpeptidation of peptides, leading to an equilibrium mixture of small and large peptides. Removal of the larger peptides results in their synthetic reappearance (Kauffman 1986).

Eigen (1971) immediately rejected such a system for, as he argued, it could not evolve (Fig. 15). There is absolutely no guarantee that a mutant enzyme E'_1 will trigger a mutant sub-network leading to the specific reinforcement of itself. Protein networks might be units of selection, but they are not units of evolution.

Dyson (1982, 1985) has shown that *if* such a network is possible (there exists at least one ensemble of peptides such as the graph of interactions

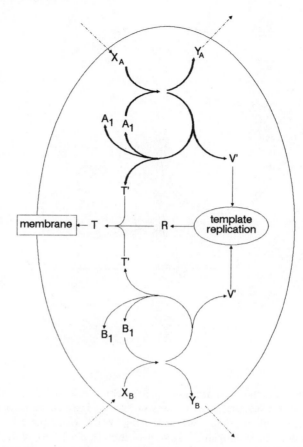

Fig. 13. Autocatalytic chemical cycles as units of selection but not as units of evolution. The figure is a simplification of Fig. 12 except for the fact that now two different autocatalytic cycles can produce the same internal materials. If one cycle (cycle A) turns round faster than the other, the intermediates of the slowly growing cycle will be lost finally from the enclosing chemoton (cf. Fig. 12). This process is irreversible.

among them includes a reflexively autocatalytic network), a spontaneous random jump in a mixture of peptides from molecular chaos (an incomplete network) to the functioning network *is* possible. In his view, such networks preceded nucleic acids, which appeared under the catalytic influence of protein enzymes, initially as parasites, later as symbionts, and ultimately they replaced the whole protein network.

Kauffman (1986) went one crucial step further and showed that such reflexively autocatalytic protein networks could in fact exist. This chapter is not the appropriate place for describing the details, but the essence is as

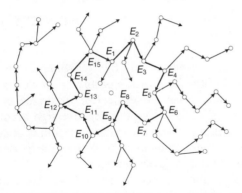

Fig. 14. A catalytic network of proteins. Each protein is multifunctional. This can give rise, under suitable circumstances, to a closed, reflexively autocatalytic loop $E_1, \ldots, E_{15}$. (Reproduced from Eigen 1971.)

follows. For combinatorial reasons, the number of possible reactions (among peptides) increases much more than the number of possible peptides as the lengths of chains is increased. From this it can be shown that there is a fair probability of the emergence of such a network in a reasonably sized peptide mixture. As proteins are more readily produced in prebiotic experiments than nucleic acids, this result favours the proteins-early view.

However, Kauffman (1986) has not solved the problem of heredity. A very recent abstract model indicates that it might also appear in *certain* kinds of networks. I will present the model in some detail because its implications are unexpected and potentially far-reaching.

Abbott (1988) starts his system with a molecular universe consisting of n chains of length N. We assume that there are only two kinds of monomers, labelled by -1 and $+1$. The population is described by the matrix S_i^a, showing the value of the ith site in the ath chain. Time is counted in discrete steps. The new value of a given site is expressed by the following equation:

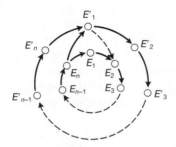

Fig. 15. 'Mutation' in a protein cycle. There is no guarantee that E'_1 will give rise to a whole mutant cycle, $E'_2, \ldots, E'_n$, and that this cycle will be selectively established. Reproduced from Eigen (1971).

$$(S_i^a)' = \text{sign}(\sum_{j=1}^n [(\sum_{a=1}^n B_{ij}S_i^a S_j^a) - B_{ij}S_i^a S_j^a]S_j^a) \qquad (27)$$

where B_{ij} is an interaction matrix with the following properties: $B_{ij} = B_{ji}$ and $B_{ii} = 0$. This means that the ith position of any molecule does not determine the ith position of any other molecule (the analogue of base-pairing is excluded). The expression following the negative sign accounts for the fact that a given chain cannot catalyse the transformation of itself. Equation (27) assumes that almost all positions in all molecules partly determine the next value of any given site in any given chain.

Simulation of the model proceeds as follows: we start with $n = n_f/2$ randomly generated chains. We apply the transformation equation (27) to this population until all chains reach a recurrent state, identical to a locally stable fixed-point of this recursion. (It should be noted that such a state corresponds to a local minimum of an effective free energy of the system; see the original reference for details.) Then a so-called food-chain, a 'peptide' of arbitrary sequence, randomly chosen but fixed for a given run, is added to the system, and eqn (27) is applied again until a new fixed-point is attained. This procedure is reiterated until the system reaches the size $n = n_f$. Then, the system is randomly split into two daughter systems, each containing $n_f/2$ chains.

To check for replication, there are two options: the regular listing and comparison of the chains present, and the application of some useful measure correlating well with population identity. It turns out that the constancy of:

$$t = (\sum_{i=1}^n \sum_{j=1}^n \sum_{a=1}^n B_{ij}S_i^a S_j^a \theta_i \theta_j)/(2N\sqrt{(nN)}) \qquad (28)$$

is a good indicator of overall population recurrence, where θ is a randomly chosen test chain. Compared with eqn (27), the above equation indicates the overall catalytic strength that a randomly chosen molecule is feeling. t is always measured before population splitting, and the daughter system, whose further fate is to be followed, is also randomly chosen from the two offspring. The computer runs show that many systems have a high degree of self-reproduction, and even some regenerative properties. This property is maintained despite the addition of considerable noise ('mutations'). Thus Abbott's model shows how multiplication and heredity can arise as emergent properties in an abstract universe without complementary template replication.

The interesting results go beyond that, however. One is able to select for increased or decreased values of t by choosing the daughter system which is closer to the required value. This directional selection effectively shifts the mean and also narrows the distribution. The system has got *hereditary variation without template replication*.

Inspection of Abbott's results reveals that the clue to this phenomenon is the compartmentation of the systems. It has the same effect on reflexively autocatalytic networks as on hypercycles: individuals with different fitness values may arise. Going back to Eigen's simplified scheme (Fig. 15) the following interpretation can be given. Alternative protein cycles correspond to alternative locally stable dynamical configurations in sequence space. Compartmentation has two effects: first, through sampling error, mutations from the attraction domain of one recurrent configuration into that of another one are possible, and secondly, populations belonging to different attraction domains can be distinguished as individuals.

However, strictly speaking, Abbott's abstract system is not a model of anything. The updating equation (27) does not correspond to any conceivable interaction in protein networks. There can be various condensation, cleavage, and exchange reactions among polypeptides, but the site-by-site exchange of amino acids, which eqn (27) would imply, is unrealistic. Although Abbott's results are interesting and promising, more plausible models in this topic would be welcome.

To conclude this section, I emphasize that the chicken–egg problem of early evolution is as acute as ever. Two decades ago, it seemed that nucleic acids were incapable of enzymatic catalysis and proteins were incapable of reproduction. The present situation is rather different (Table 2). Even for proteins, one should not forget that 'there has not yet been a generally accepted synthetic scheme demonstrated for the prebiological formation of peptides' (Bada and Miller 1987). As to nucleic acids, their simpler analogues, having something simpler instead of ribose in the backbone, thereby possibly gaining a + sign in each row above, are the only real hope (Orgel 1986).

9. RIBOZYME GENERATION BY ARTIFICIAL EVOLUTION: A PROPOSAL

In Table 2 nucleic acids have been assigned to a single + in the column of enzymatic function. This is attributable to the fact that, despite speculations (Woese 1967; Crick 1968; Orgel 1968; White 1976; Korányi and Gánti 1981),

Table 2
The current status of the chicken–egg problem in evolutionary molecular biology

	Enzymatic function	Replicative ability	Spontaneous prebiotic formation
Proteins	+ +	+?	+
Nucleic acids	+	+ +	Analogues?

all reported ribozymic activities are associated with reactions transforming other nucleic acids. Recent, educated speculations suggest that RNAs may be capable of forming hydrophobic pockets, general acid-base catalysis, and the formation of covalent intermediates (Cech and Bass 1986).

Nevertheless, a systematic method for testing ribozymic capabilities fairly generally would be a benefit to the field. I present here a revised version of my former suggestions (Szathmáry 1984, in press), following some necessary background material.

Following Pauling's ideas, it has become clear that enzymatic catalysis rests on the fact that the transition complex is better bound than the substrates, thus the activation energy is reduced and the reaction rate is enhanced (Lienhard 1973; see Fig. 16).

This knowledge has been extensively used in the production of catalytic antibodies. Monoclonal antibodies were generated against a *stable* chemical analogue of the transition complex of the desired enzymatic reaction, and it was found that the antibody was capable of acting as an enzyme (Napper *et al.* 1987; Schultz 1988). This method is evolutionary in the sense that it uses the variation–selection processes of the immune system.

Coming to ribozymes, the reaction and a stable transition complex analogue have to be chosen accordingly. One should select, in an evolutionary process, for RNAs which bind the analogue strongly.

However, the brute force application of Spiegelman's (1970) experimental system leads into an impasse. In that scenario the sole criterion of survival is replicative ability. It would be foolish to expect that the best-replicating RNAs are nicely pre-adapted for enzymatic activity. Somehow the RNA population should be made to undergo a certain number of average (low) number of replication rounds, and selection for the binding efficiency to the transition complex analogue should be intermittent. Binding efficiency could

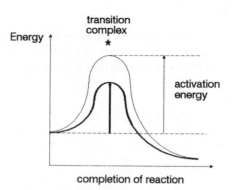

Fig. 16. Energy profiles of uncatalysed and catalysed reactions. The transition complex is designated by an asterisk. The enzyme lowers the activation energy by binding the transition complex.

be measured on an affinity chromatography column, and only the best-binding RNAs should be propagated.

Eigen's (1986) evolutionary reactor seems ideally suited for such a purpose. Still under development, it enables one to make a systematic and effort-saving search through the sequence space (Maynard Smith 1970) of replicating RNA molecules. One can start with an arbitrary, homogeneous RNA population. Controlled circumstances in the reactor make it possible to adjust the average mutational distance of the offspring molecules. Members of this, now heterogeneous, population can be cloned via a reversed transcribed DNA intermediate, to make use of the much higher fidelity of DNA replication (M. Eigen personal communication). The amplified sequences can be re-transcribed into RNAs again. It is at this stage where chromatographic selection should take place. More detailed searches in sequence space should be done around sequences which show a higher affinity towards the transition complex analogue than the others do. I predict that through the application of this experimental scheme, production of various ribozymes should be possible. (It should be noted that Eigen (1986) raises the possibility of the production of new *protein* enzymes which involves some testing of *translated* gene products.)

The realization of this idea could lead to a breakthrough, at least on the theoretical level. As an example, think of the origin of translation. One recently revived idea is the stereochemical hypothesis. According to Shimizu (1982, 1987) and Yoneda *et al.* (1985), there is a lock and key relationship between the anticodon–discriminator base complex (the so-called C4N complex) and the cognate amino acid. Whether this idea, or some other alternative, resting on the presumed amino acid recognition capabilities of tRNA analogues, is promising or not could be tested in the system outlined above. In this case, one should select for the specific binding of RNAs to amino acids.

It is interesting to note that Kinjo *et al.* (1986) have constructed a 'model' primitive tRNA which the Qß replicase accepts as a template and replicates. Further work could be possible following the proposed scenario.

ACKNOWLEDGEMENTS

The author wishes to thank the following for advice and discussions: Michael Bulmer, Jack Corliss, László Demeter, Manfred Eigen, Tibor Gánti, Fred Hopf, Pál Juhász-Nagy, Géza Meszéna, István Molnár, István Scheuring, Karl Sigmund, Mária Ujhelyi, László Vekerdi, and Gábor Vida. Special thanks are due to John Maynard Smith for acting as a supervisor during a six-month visit to the University of Sussex granted by the SOROS Foundation (New York).

REFERENCES

Abbott, L. F. (1988). A model of autocatalytic replication. *J. Mol. Evol.* **27**, 114–20.

Alberts, P. *et al.* (1983). *Molecular biology of the cell.* Garland Publishing, New York.

Bada, J. L. and Miller, S. L. (1987). Racemization and the origin of optically active organic compounds in living organisms. *BioSystems* **20**, 21–6.

Bauer, E. (1967). *Theoretical biology*, (in Hungarian). Akadémiai Kiadó, Budapest.

Biebricher, C. K., Eigen, M., and Gardiner, W. C. (1983). Kinetics of RNA replication. *Biochemistry* **22**, 2544–59.

Biebricher, C. K., Eigen, M., and Gardiner, W. C. (1985). Kinetics of RNA replication: competition and selection among self-replicating RNA species. *Biochemistry* **24**, 6550–60.

Bloch, D. P., McArthur, B., and Mirrop, S. (1985). tRNA–rRNA sequence homologies: evidence for an ancient modular format shared by tRNAs and rRNAs. *BioSystems* **17**, 209–25.

Blow, J. J., Dilworth, S. M., Dingwall, C., Mills, A. D., and Laskey, R. A. (1987). Chromosome replication in cell-free systems from *Xenopus* eggs. *Phil. Trans. R. Soc. Lond. B* **317**, 483–94.

Bresch, C., Niesert, U., and Harnasch, D. (1980). Hypercycles, parasites, and packages. *J. Theor. Biol.* **85**, 399–405.

Bulmer, M. J. (1988). Evolutionary aspects of protein synthesis. In *Oxford Surveys in evolutionary biology*, Vol. 5 (ed. P. H. Harvey and L. Partridge), pp. 1–40. Oxford University Press.

Buss, L. (1988). *The Evolution of Individuality.* Princeton University Press, Princeton, New Jersey.

Cavalier-Smith, T. (ed.) (1985). *The evolution of genome size.* Wiley, Chichester.

—— (1987). The simultaneous symbiotic origin of mitochondria chloroplasts, and microbodies. *Ann. N. Y. Acad. Sci.* **503**, 55–71.

Cech, T. and Bass, B. L. (1986). Biological catalysis by RNA. *Ann. Rev. Biochem.* **55**, 599–629.

Crick, F. H. C. (1968). The origin of the genetic code. *J. Mol. Biol.* **38**, 367–79.

—— (1982). *Life itself—its origin and nature.* Futura, London.

——, Brenner, S., Klug, A., and Pieczenik, G. (1976). A speculation on the origin of protein synthesis. *Origins of Life* **7**, 389–97.

Dawkins, R. (1982). *The extended phenotype.* W. H. Freeman, Oxford.

Dyson, F. (1982). A model for the origin of life. *J. Mol. Evol.* **18**, 344–50.

Dyson, F. J. (1985). *Origins of life.* Cambridge University Press.

Eigen, M. (1971). Selforganization of matter and the evolution of biological macromolecules. *Naturwiss.* **58**, 465–523.

—— (1986). The physics of molecular evolution. *Chemica Scripta* **26B**, 13–26.

—— and Schuster, P. (1977). The Hypercycle. A principle of natural self-organization. Part A: emergence of the hypercycle. *Naturwiss.* **64**, 541–65.

—— and —— (1978*a*). The hypercycle. A principle of natural self-organization. Part B: the abstract hypercycle. *Naturwiss.* **65**, 7–41.

—— and —— (1978*b*). The hypercycle. A principle of natural self-organization. Part C: the realistic hypercycle. *Naturwiss.* **65**, 341–569.

—— and —— (1982). Stages of emerging life—five principles of early organization. *J. Mol. Evol.* **19**, 47–61.

——, Gardiner, W. C. Jr., and Schuster, P. (1980). Hypercycles and compartments. *J. Theor. Biol.* **85**, 409–11.

——, Schuster, P., Gardiner, W., and Winkler-Oswatitsch, R. (1981). The origin of genetic information. *Sci. Am.* **244(4)**, 78–94.

Epstein, I. R. (1979). Competitive coexistence of self-reproducing macromolecules. *J. Theor. Biol.* **78**, 271–98.

Gánti, T. (1966). *Revolution in life research*, (in Hungarian). Gondolat, Budapest.

—— (1971). *The principle of life*, (in Hungarian). Gondolat, Budapest.

—— (1975). Organization of chemical reactions into dividing and metabolizing units: the chemotons. *BioSystems* **7**, 189–95.

—— (1978). Self-regulation and self-organization of molecular supersystems, (in Hungarian). In *Biological Regulation* (ed. G. Csaba), pp. 37–73. Medicina Könyvkiadó, Budapest.

—— (1979). *A theory of biochemical supersystems and its application to problems of natural and artificial biogenesis.* Akadémiai Kiadó, Budapest/University Park Press, Baltimore.

—— (1984). *Chemoton theory I. Theoretical foundations of fluid machineries*, (in Hungarian). OMIKK, Budapest.

—— (1987). *The principle of life.* OMIKK, Budapest.

Gause, G. F. (1934). *The struggle for existence.* Williams & Wilkins, Baltimore.

Haken, H. (1983). *Advanced synergetics.* Springer, Berlin.

Hull, D. L. (1980). Individuality and selection. *Ann. Rev. Ecol. Syst.* **11**, 311–32.

Jones, B. L., Enns, R. H., and Rangnekar, S. S. (1976). On the theory of selection of coupled macromolecular systems. *Bull. Math. Biol.* **38**, 15–28.

Kauffman, S. (1986). Autocatalytic sets of proteins. *J. Theor. Biol.* **119**, 1–24.

King, C. C. (1982). A model for the development of genetic translation. *Origins of Life* **12**, 405–25.

King, G. A. M. (1981). Growth of a hypercycle and comparison with conventional autocatalysis. *BioSystems* **13**, 225–34.

Kinjo, M., Hasegawa, T., Nagano, K., Ishikura, H., and Ishigami, M. (1986). Enzymatic synthesis and some properties of a model primitive tRNA. *J. Mol. Evol.* **23**, 320–7.

Koŕanyi, P. and Ganti, T. (1981). Coenzymes as remnants of ancient enzyme ribonucleic acid molecules, (in Hungarian). *Biológia* **29**, 107–24.

Küppers, B.-O (1983). *Molecular theory of evolution.* Springer, Berlin.

Lienhard, G. E. (1973). Enzymatic catalysis and transition-state theory. *Science* **180**, 149–54.

Margulis, L. (1981). *Symbiosis in cell evolution.* W. H. Freeman, San Francisco.

Maynard Smith, J. (1970). Natural selection and the concept of a protein space. *Nature* **225**, 563–4.

—— (1978). *The evolution of sex.* Cambridge University Press.

—— (1979). Hypercyles and the origin of life. *Nature* **280**, 445–6.

—— (1983). Models of evolution. *Proc. R. Soc. Lond. B* **219**, 315–25.

—— (1986). *The problems of biology.* Oxford University Press.

—— (1987). How to model evolution. In *The latest on the best. Essays on evolution and optimality* (ed. J. Dupré), pp. 119–31. MIT Press, Cambridge, Massachussetts.

Michod, R. (1983). Population biology of the first replicators: on the origin of genotype, phenotype and organism. *Amer. Zool.* **23**, 5–14.

—— (1984). Contraints on adaptation, with special reference to social behavior. In *A New Ecology* (ed. P. W. Price, C. N. Slobodchikoff, and W. S. Gaud), pp. 253–78. Wiley, New York.

Napper, A. D., Benkovic, S. J., Tramontano, A., and Lerner, R. A. (1987). A stereospecific cyclization by an antibody. *Science* **237**, 1041–3.

Niesert, U., Harnasch, D., and Bresch, C. (1981). Origin of life between Scylla and Charybdis. *J. Mol. Evol.* **17**, 348–53.

Ohno, S. (1987). Evolution from primordial oligomeric repeats to modern coding sequences. *J. Mol. Evol.* **25**, 325–9.

Orgel, L. E. (1968). Evolution of the genetic apparatus. *J. Mol. Biol.* **38**, 381–93.

—— (1986). RNA catalysis and the origins of life. *J. Theor. Biol.* **123**, 127–49.

Roberts, J. M. and Weintraub, H. (1988). *Cis*-acting negative control of DNA replication in eukaryotic cells. *Cell* **52**, 397–404.

Rodriguez-Vargas and Schuster, P. (1984). The dynamics of catalytic hypercycles—a stochastic simulation. In *Stochastic phenomena and chaotic behaviour in complex systems* (ed. P. Schuster), pp. 208–19. Springer, Berlin.

Schrödinger, E. (1967). *What is life? & mind and matter.* Cambridge University Press.

Schultz, P. G. (1988). The interplay between chemistry and biology in the design of enzymatic catalysts. *Science* **240**, 426–32.

Shimizu, M. (1982). Molecular basis for the genetic code. *J. Mol. Evol.* **18**, 297–303.

—— (1987). Precise ultraviolet absorbance study of the interaction of amino acids and mononucleosides in aqueous solution. *Biophys. Chem.* **28**, 169–74.

Spiegelman, S. (1970). Extracellular evolution of replicating molecules. In *The neurosciences: second study program* (ed. F. O. Schmitt), pp. 927–45. Rockefeller University Press, New York.

Swetina, J. and Schuster, P. (1982). Self-replication with errors. A model for polynucleotide replication. *Biophys. Chem.* **16**, 329–45.

Szathmáry, E. (1984). The roots of individual organization, (in Hungarian). In *Evolution IV. Frontiers of evolutionary research*, (ed. G. Vida), pp. 37–157. Natura, Budapest.

—— (1986a). The eukaryotic cell as an information integrator. *Endocyt. C. Res.* **3**, 113–32.

—— (1986b). Some remarks on hypercycles and the stochastic corrector model. *Endocyt. C. Res.* **3**, 337–39.

—— (1988). A hypercyclic illusion. *J. Theor. Biol.* **134**, 561–3.

—— (1989). The integration of genetic information and the emergence of novel evolutionary units. *Trends Ecol. Evol.* **4**, 200–4.

—— (in press). RNA worlds in test-tubes and protocells. In *Organizational constraints on the dynamics of evolution* (ed. G. Vida and J. Maynard Smith). Manchester University Press.

—— and Demeter, L. (1986). Dynamical co-existential interactions in prebiotic systems II. Dynamics of replicators with an aspecific mutual aid (in Hungarian). *Biológia* **34**, 39–64.

—— and —— (1987). Group selection of early replicators and the origin of life. *J. Theor. Biol.* **128**, 463–86.

—— and Gladkih, I. (1989). Coxistence of non-enzymatically replicating templates. *J. Theor. Biol.* **138**, 55–8.

——, Kotsis, M., and Scheuring, I. (1988). Limits of hyperbolic growth and selection in molecular and biological populations. In *Mathematical ecology* (ed. T. G. Hallam, L. J. Gross, and S. A. Levin), pp. 45–67. World Scientific, Singapore.

Ujhelyi, M. (1983). The possible common origin of tRNA and 5S rRNA. *Z. Naturforsch.* **38,** 501–4.

von Kiedrowski, G. (1986). Ein selbst-replizierendes hexadesoxynukleotid. *Angew. Chem.* **98,** 932–4.

Weiner, A. M. and Maizels, N. (1987). tRNA-like structures tag the 3' ends of genomic RNA molecules for replication: implications for the origin of protein synthesis. *Proc. Natl Acad. Sci. USA* **84,** 7383–7.

White, H. B. (1976). Coenzymes as fossils of an earlier metabolic stage. *J. Mol. Evol.* **7,,** 101–4.

Wilson, D. S. (1983). The group selection controversy: history and current status. *Ann. Rev. Ecol. Syst.* **14,** 159–87.

Woese, C. R. (1967). *The genetic code: The molecular basis for genetic expression.* Harper & Row, New York.

Yoneda, S., Shimizu, M., Fujii, S., and Go, N. (1985). Induced fitting between a complex of four nucleotides and the cognate amino acid. *Biophys Chem.* **23,** 91–104.

Zielinski, W. S. and Orgel, L. E. (1987). Autocatalytic synthesis of a tetranucleotide analogue. *Nature* **327,** 346–7.

Population genetics at the DNA level: a review of the contribution of restriction enzyme studies

ANDREW J. LEIGH BROWN

1. INTRODUCTION—RESTRICTION ENZYMES: WHY THEY ARE USED

1.1 General

Population genetics is one of the branches of evolutionary biology that can, in principle, be applied to all living systems. It deals with the forces which act on the hereditary material and determine the extent to which succeeding generations retain the features of their ancestors. These forces were identified early on and can be summarized simply: mutation, selection, migration, and sampling, or drift. All processes which give rise to heritable change in the constitution of natural populations, i.e. evolution, can be allocated to one of these divisions. If their values are all equal to zero, a population is not expected to evolve.

Throughout most of this century, the conceptual organization of the field of population genetics, and indeed of genetics in general, has remained as it was at its inception. Heritable variation results from the presence of two or more phenotypically distinguishable allelomorphs which segregate in a genetic cross, at loci which can be separated by recombination. This simple Mendelian view of genetics has been tremendously powerful. It enabled the development of the most sophisticated and rigorous body of theory in the whole of biology, which has the capacity to describe the genetic outcome for a population, in probabilistic terms, under a wide range of different circumstances. However, within the last 15 years, a revolution has occurred in genetics and the consequences for population genetics are profound.

For all known organisms (with the possible exception of the scrapie agent Weissman 1989), the hereditary information is contained in a nucleotide sequence, be it only a few hundred base-pairs long as in the case of the plant viroids (Riesner and Gross 1985), or more than 10^9 base-pairs as in vertebrates and angiosperms. With the publication in 1977 of two methods for the rapid determination of nucleotide sequences (Maxam and Gilbert 1977; Sanger et al. 1977), it became possible to obtain data on heritable variation in all organisms, including viruses. We now have the capability of detecting *all* such variation, and for any sequence in which we might be interested. Thus, for the first time, the methods of population genetics became applicable in practice to all forms of life. Indeed, the very first organism for which the entire genome was sequenced from several individuals and is available for comparative purposes was a virus: HIV–1, the AIDS virus (Ratner *et al.* 1985; Sanchez-Pescador *et al.* 1985; Wain Hobson *et al.* 1985). However, the nature of the data thus obtained is significantly different to that on which population genetic theory was based. A Mendelian gene was a single indivisible unit. Now we know not only that recombination can occur within genes, generating new allelic classes (Strobeck and Morgan 1978), but also that the genomic DNA sequence for most genes is split into exons, which contain the coding information and non-coding introns. The rates of evolution of introns and exons are often quite different. We know that some genes are repeated many times in a genome, and that different copies can exchange information; and we know of 'jumping genes'— transposable genetic elements whose presence is undetectable at the phenotypic level unless they cause a mutation. What impact has the new, nucleotide level, view of the gene had on population genetics in the last 12 years?

For several reasons, the actual determination of nucleotide sequence has, as yet, not been a feature of many population studies, with the notable exceptions of Kreitman's analysis of *Adh* sequences in *Drosophila melanogaster* (Kreitman 1983) and of some recent studies on HIV-1 (Hahn *et al.* 1986). However, another tool of molecular genetics, the restriction enzyme, has been adopted extensively. Although restriction enzymes are much less sensitive than nucleotide sequencing in detecting sequence changes, their widespread adoption has begun to show to what extent the new type of data can be fitted in to the existing framework of population genetics, where its greatest value may lie, and where it might raise more questions than answers. In this chapter I shall review the uses to which restriction enzymes have been put in population genetics and some of the results that have been obtained. I will attempt to assess the value of this new molecular tool and the implications of the new methodology for the field in general.

1.2 Restriction enzymes

Restriction enzymes are a group of DNA-binding proteins which bind to short nucleotide sequences, usually 4–6 base-pairs in length. Each enzyme recognizes a specific target sequence, and alteration of a single base within that target sequence is sufficient to prevent recognition and binding. After binding the enzyme cleaves both strands of the DNA molecule leaving blunt ends or short single-stranded protrusions, known as 'sticky' ends. Some

Table 1

Commonly used restriction enzymes, their recognition sequences, and sites of cleavage

Enzyme	Recognition seq.		Cleavage sites	
			After cleavage	
Eco R I	5′ GAATTC	3′	5′ G	3′
	3′ CTTAAG	5′	3′ CTTAA	5′
			After cleavage	
Bam H I	5′ GGATCC	3′	5′ G	3′
	3′ CCTAGG	5′	3′ CCTAG	5′
Pst I	5′ CTGCAG	3′	5′ CTGCA	3′
	3′ GACGTC	5′	3′ G	5′
	C A			
Hin c II	5′ GT or or AC		5′ GTC	5′
	T G		3′ CAG	5′
			or	
	A C			
	3′ CA or or TG		5′ GTT	5′
	G T		3′ CAA	3′
Sau 3A I	5′ GATC	3′	5′	3′
	3′ CTAG	5′	3′ CTAG	5′

examples of commonly used restriction enzymes are given in Table 1 with their recognition sequences. The nature of the termini after cleavage is also shown. The fact that such an enzyme will generate the same sticky end from cleavage of any DNA molecule was exploited by Boyer and Cohen and their colleagues to ligate together two unrelated pieces of DNA and initiate the recombinant DNA revolution (Cohen *et al.* 1973). For the purposes of population geneticists, the exact nature of the termini after cleavage is not usually important, and the significant features of restriction enzymes are the site specificity of their binding to DNA and the coincidence of binding and cleavage.

The enzymes used in population studies, indeed most of those used in molecular genetics, belong to a class known as type II. They can be isolated from a wide variety of prokaryotes, are usually plasmid encoded and require only Mg^{2+} ions for activity (reviewed by Bickle 1987; Modrich and Roberts 1982). Two other classes of restriction enzyme are known, each of which also requires ATP at least for cleavage. These are the type I and type III systems which are quite different in biochemical terms although their biological role is similar (reviewed by Yuan 1981). All restriction enzymes form part of 'restriction and modification' systems, whose function appears to be to protect the prokaryotic cell from infection by bacteriophages. This is achieved by modification (usually involving methylation of particular bases) at specific short sequences of DNA. The second component is the cleavage of all DNA molecules which are not modified at that sequence. In consequence, any bacteriophage which has grown on a particular host strain will be *restricted* to that strain because it will share its pattern of modification. Attempts to infect a different host strain with a different restriction and modification system will be unsuccessful because the bacteriophage DNA will be recognized as foreign due to the lack of methylation at the appropriate sites and will be cleaved by the restriction enzyme.

In type II systems, DNA cleavage and modification occurs as a quite distinct reaction and is carried out by two separate enzymes each of which can, in principle, be purified on its own. In type I and type III systems, both reactions are carried out by a single multifunctional enzyme, and cannot therefore be separated. Although up to 200 type II restriction enzymes are known and many are commercially available, only a relatively small number are used in population studies. I shall discuss the basis for the use of restriction enzymes in population genetics and the techniques involved before going on to consider in detail the results which have been obtained.

1.3 Use of restriction enzymes in population genetics

Restriction enzymes have been used to approach several distinct, even unrelated problems in population genetics (Table 2). One of the first applications was as a means of estimating genetic variation and rate of evolution at the nucleotide level. The approach was straightforward and was first applied to the relatively simple situation of the animal mitochondrial genome (Avise *et al.* 1979; Shah and Langley 1979). The DNA, extracted from each of the groups of organisms under study, was digested with a restriction enzyme, and the fragment pattern obtained after gel electrophoresis (see Section 1.4) was compared. If there are few differences in such a fragment pattern, it can be assumed that fragments migrating at the same speed (i.e. fragments of the same size), are truly homologous. The nucleotide divergence was then estimated on the basis of the proportion of differences in the fragment pattern (Nei and Li 1979). When there are several differences in the fragment pattern, it is necessary to establish the identity of each of the fragments in the DNA samples, because there is then a non-trivial probability of non-homologous fragments migrating at the same speed through convergence. The identity of each fragment can readily be determined by restriction mapping. In this process the locations of all cleavage sites of two

Table 2

Major areas of use of restriction enzymes in population genetics

Area	Material	Methods	Alternative approaches
1. Estimation of nucleotide substitution rate and diversity	mtDNA Nuclear DNA	EtBr† staining Terminal labelling Southern transfer	(Allozymes)* Cloning/ sequencing
2. Analysis of population structure	mtDNA	EtBr staining Terminal labelling Southern transfer	Allozymes
3. Evolution of repetitive DNA sequences	Nuclear DNA	Southern transfer	Cloning/ sequencing
4. Genome rearrangements (transposition)	Nuclear DNA	Southern transfer	*In situ* hybridization
5. Linkage disequilibrium	Nuclear DNA	Southern transfer	Allozymes

*Estimation of nucleotide diversity from allozyme data is very indirect.
†Ethidium bromide.

or more enzymes in the DNA molecules are established by recording the fragment pattern resulting from single-enzyme and double-enzyme digests in all combinations. The restriction maps of the samples are then compared and the nucleotide divergence is estimated from the proportion of sites which are conserved (Engels 1981; Ewens *et al.* 1981; Hudson *et al.* 1982; Nei and Tajima 1981, 1983).

In a larger molecule, such as chromosomal DNA, the specific region of interest must first be identified. This is achieved by allowing an homologous DNA sequence which has been radioactively labelled to base-pair with the DNA fragments after digestion—a procedure referred to here as Southern transfer-hybridization (see below). The data collected from such surveys consist of an array of binary characters, i.e. the presence or absence in each individual of each restriction site detected. The absence of a site from one or more individuals is interpreted as indicating a single nucleotide substitution within the recognition site in that individual, and the gain of a new site represents the reverse process. Thus a relatively direct interpretation of the frequency of nucleotide substitutions can be made from restriction site data.

Genetic variability represents such a corner-stone of evolutionary thinking that its quantification has a long history (reviewed by Lewontin 1974). Until the advent of protein electrophoresis, there was no method which could detect genes or gene products in the absence of pre-existing variation. Electrophoretic surveys of enzyme variation made it possible for the first time to say how many genes were monomorphic and how many polymorphic. Electrophoretic studies also allowed the detection of variation without regard to its phenotypic effect. These important advances are met equally well by restriction enzymes which have the further advantage that the data can be interpreted directly in terms of numbers of nucleotide substitutions. In addition, whereas enzyme electrophoresis was constrained to studying protein-coding regions, restriction enzyme analysis can be directed towards both coding and non-coding regions, of which the latter represent the vast majority of the DNA in eukaryotes, and to the study of extranuclear genomes. We shall see below how important these advantages have been.

One of the first evolutionary observations to come out of restriction enzyme studies of the rate of nucleotide substitution was that the mitochondrial genome of animals, which was now accessible to evolutionary studies for the first time, evolved much more rapidly than nuclear genes. The mitochondrial genome is transmitted in a clonal fashion through the maternal lineage. These two factors were fundamental to the development of mitochondrial DNA (mtDNA) studies in population structure. The absence of recombination means that each mtDNA molecule can be treated as a single, multi-allelic locus, by analogy with the genetic studies of MHC loci in mammals where the term 'haplotype' was first introduced to describe the linked array of alleles. Because of the large number of possible allelic combinations when several polymorphic restriction sites have been found, phylogenetic analysis has often been used to determine relationships between populations rather than genetic distance estimates based on gene frequencies.

Another area of application of restriction enzymes has also developed because of the possibility of accessing previously inaccessible segments of the genome. Many eukaryotic genomes are characterized by the presence of a large fraction of repetitive DNA sequences, often tandemly arrayed. These are resistant to genetic analysis because of their redundancy, loss of function of individual copies being phenotypically undetectable. However, they were some of the first genes to be studied using molecular techniques. Although some of the arrays of repetitive sequences encode vital functions, the histone genes and ribosomal RNA genes (rDNA) for example, others such as simple satellite sequences remain without any function ascribed to them. With restriction enzymes it was possible to distinguish sequence differences between members of such gene families and to make comparisons within and between species, with startling results.

Two more fields of use of restriction enzymes have developed from the initial studies on the levels of genetic variability (Table 2). The first concerns the analysis of linkage disequilibrium between polymorphic sites and the second the analysis of variation other than that due to nucleotide substitution. The former represents a direct extension of studies which had previously been carried out using enzyme polymorphisms in many natural populations (Langley 1977). However, allozyme data suffer from certain disadvantages, in particular the presence within each electrophoretic class of more than one nucleotide sequence. This is a much less serious problem for polymorphic restriction sites as the probability of more than one nucleotide polymorphism within a 6 base sequence is extremely small. The latter subject is specific to the analysis of DNA sequences. Even the earliest studies using restriction enzymes to analyse nuclear genes in *Drosophila melanogaster* discovered a high level of variation due to insertion and deletion of large segments of DNA. These were subsequently identified as being transposable genetic elements which are well nigh ubiquitous in prokaryotic and eukaryotic genomes alike. Such insertions give rise to major changes in the restriction map of a region without necessarily removing any restriction sites. They are screened for by looking for alterations in the pattern of restriction fragments that affect more than one enzyme in the same region. As different restriction enzymes recognize quite different target sequences (Table 1), the only explanation of such observations is that an insertion (or deletion) of DNA has occurred. Although variation due to the mobility of transposable elements can also be studied by *in situ* hybridization using cloned sequences of particular transposable elements, restriction enzymes are a valuable way of analysing variation arising from this source as they will detect all such events in the studied region, regardless of the identity of the inserted DNA or of whether they give rise to any phenotypic effect.

1.4 Methodology

For all types of study it is necessary to render the restriction fragments obtained visible and identifiable. In the first instance, restriction fragments are separated according to size by electrophoresis in agarose or occasionally in acrylamide gels. There is more than one method whereby the fragments can be visualized and the method chosen will usually depend on the nature of the starting material.

When it is possible to obtain the DNA molecule under study purified in sufficient quantity, digestion with restriction enzymes may yield fragments which can be visualized by staining with the fluorescent intercalating dye ethidium bromide. This approach was used in early studies of mitochondrial DNA molecules which, in animals, are small covalently closed supercoiled circles that can be purified away from genomic DNA on isopycnic (equilibrium density) caesium chloride–ethidium bromide gradients in an ultracentrifuge. However, ethidium bromide staining is not very sensitive and can only reliably be used to detect DNA fragments greater than 500 base-pairs in length. An autoradiographic method can be used to detect fragments of mtDNA below this size. This involves the addition of ^{32}P to the termini of the fragments. It can be achieved by filling in the sticky ends with appropriate ^{32}P-labelled nucleotides: thus for *Eco*RI digested DNA either ^{32}PdATP or ^{32}PTTP could be used, while for *Sau*3AI (and *Bam*HI) fragments any labelled deoxynucleotide is appropriate (see Table 1). The reaction is catalysed by DNA Polymerase I (Klenow fragment).

Enzymes which leave flush or blunt ends, such as *Hinc*II (Table 1), cannot be labelled in this way. In such cases it is necessary to dephosphorylate the 5′ terminal base and replace the phosphate using ^{32}P-γ-ATP and polynucleotide kinase.

As mentioned earlier, these methods require the DNA sequences under study to be purified. For many evolutionary applications, the restriction fragments under study are not purified and may form a very small component of the DNA sample. In this case it is necessary to use Southern transfer-hybridization to detect specific fragments (Southern 1975; Jeffreys and Flavell 1977; Wahl *et al.* 1979). A digest of total genomic DNA is performed and separated in an agarose gel as before. The DNA fragments are denatured and transferred to a membrane (nylon membranes have replaced nitrocellu-lose in this application), by capillary blotting, under vacuum or by an electric field (electro-blotting). The membrane retains a semi-permanent record of the DNA separation and specific restriction fragments can be detected by hybridization with a cloned probe for the region which has previously been labelled with ^{32}P, followed by autoradiography. Appropriate labelling meth-ods include nick translation (Rigby *et al.* 1977), randomly primed synthesis, (Feinberg and Vogelstein 1984) and *in vitro* transcription (Church and Gilbert 1984). The reader is referred to a laboratory manual for details of these and other technical procedures (e.g. Berger and Kimmel 1987). Use of Southern transfer-hybridization provides a positive identification of the fragments visualized by autoradiography. Under standard procedures they must share a high level of sequence homology with the probe. Information on variation at several different sites over a large region can be built up by the successive re-hybridization of the filter with probes made from adjacent fragments, with the label being washed off the filter after each exposure.

Although restriction enzymes are a very flexible tool, there are technical limitations in their application. The resolution of a standard agarose gel system is of the order of 2–5 per cent. Thus, when screening fragments about 10 kilobases (kb) in size it would not be possible reliably to detect differences of less than 200–500 base-pairs. Secondly, the number of nucleo-tides which has been screened in most surveys is small, as each site only represents six nucleotides at most. A large number of digests are required to screen more than 200 base-pairs, although Kreitman and Aguade (1986*a*) have shown how adaptations to the standard methodology can increase the number of nucleotides screened and still permit screening of up to 80 chromosomes. A great deal of genetic variation can be uncovered in standard restriction enzyme surveys, but it is far from a complete description of the variation present at any particular gene. To that extent such surveys represent a half-way stage before the wider adoption of rapid or even automated sequencing systems. Nevertheless, the impact on the field has already been considerable.

2. RESTRICTION ENZYMES—HOW THEY ARE USED

2.1 Restriction enzymes in the analysis of mitochondrial DNA

Mitochondrial DNA (mtDNA) evolution has recently been reviewed elsewhere (Avise 1986; Avise et al. 1987; Moritz et al. 1987; Palmer 1985; Wilson et al. 1985; Harrison 1989). I will therefore focus on more recent developments and on issues which are of wider interest.

There are three major differences between mtDNA and nuclear DNA that are important in an evolutionary context: the exclusively maternal mode of transmission; the complete lack of recombination; and the very high rate of nucleotide substitution. The mitochondrial genome shows a clonal pattern of inheritance. It should be borne in mind, however, that recombination is highly suppressed in some regions of the nuclear genome and the evolution of these regions can be viewed in the same way. The best known case is that of the Y chromosome: crossing-over is completely suppressed over most of its length and, in consequence, it is inherited as a single unit, paternally transmitted. In addition it is known that in centromeric and telomeric regions of *Drosophila* chromosomes, recombination is also suppressed. In this case it is not only heterochromatic but also euchromatic DNA that is affected. The population genetic consequences are discussed in greater detail in a later section. Nevertheless, in conjunction with the ease of obtaining mtDNA (molecular cloning was often not required) and the high rate of nucleotide substitution, clonal inheritance has meant that evolutionary change in mtDNA molecules has been relatively easy to document and often can be interpreted in a fairly straightforward way.

In recent years, one of the questions to which much attention has been given is whether the pattern of evolution revealed by mtDNA converges on that obtained from studies of nuclear DNA sequences. The issue arises from the more basic observation that, in general, geographically isolated populations of many species can be distinguished on the basis of their mtDNA restriction maps (Avise et al. 1987; De Salle et al. 1987; Tegelstrom 1987; De Salle and Templeton 1988; Nelson et al. 1987; Shields and Wilson 1987; MacNeil and Strobeck 1987; Ashley and Wills 1987; Moriwaki et al. 1984; Baba-Aissa et al. 1988). However, there are several cases among those cited above where it has been shown that the distribution of mtDNA genomes bearing characteristic restriction sites may not precisely coincide with biological species boundaries, although characters encoded by the nuclear genome (diagnostic morphological characters, chromosome inversions, enzyme variants, etc.) do, almost without exception. Observations of this type have been made on *Mus, Peromyscus, Clethrionomys, Rana,* and in two species groups of *Drosophila* (Ferris et al. 1983; Avise 1986; Avise et al. 1983; Tegelstrom 1987; Spolsky and Uzzell 1984; Solignac et al. 1986; Powell 1983). Interestingly, quite different conclusions on the origin of the discrepancies have been reached in different cases. On the one hand, Avise and colleagues (Avise 1986; Avise *et al*. 1984; Neigel and Avise 1986) have provided an argument on the probability of extinction of individual mtDNA clones derived from the theory of branching processes developed by Haldane (1927). In large populations where the variance of litter size is not large relative to the mean, mtDNA restriction maps which were present in the ancestral population before the speciation event could persist within the sexually (and even geographically) isolated populations for some considerable time afterwards. They have applied these conclusions to their own observations on *Peromyscus* (Avise *et al*. 1983), where they would appear to provide the most convincing explanation of their observations that the mtDNA of some populations of *P. maniculatus* is more closely related to that of certain populations of another species, *P. polionotus*. These populations are not sympatric and there is no evidence that hybridization has been involved.

In many other cases, a quite different explanation is offered for apparently similar observations. This is best exemplified by the case described by Ferris *et al*. (1983) (reviewed by Wilson *et al*. 1985). A hybrid zone exists over much of Europe between the two subspecies *Mus musculus musculus* and *M.m. domesticus*, with the mitochondrial restriction maps of mice caught on either side of the zone corresponding to their subspecies as determined by nuclearly encoded genes (Sage and Wilson, cited in Wilson *et al*. 1985). However, specimens that were caught from north of the hybrid zone in Denmark which were *M.m. musculus* from their nuclear genome, contained mtDNA restriction maps characteristic of *M.m. domesticus*. In addition, all mice caught in Sweden, where only *musculus* is found, share the *domesticus* mtDNA maps.

It would be quite unexpected for the retention of ancestral mtDNA types, on a probabilistic basis as discussed by Neigel and Avise (1986) to give rise to such large coherent populations and not to occur elsewhere within *M.m. musculus*. The explanation favoured by Wilson and colleagues, that a rare escaper occurred from the usual inviability associated with subspecific hybrids which led to the 'introgression' of the *domesticus* mtDNA into a predominantly *musculus* population. They suggested that this event was associated with the founding of the northern Danish and Swedish populations in post-glacial times. The possibility of a reduced level of intraspecific competition at such a time could explain the difference between that population and their observations across the hybrid zone elsewhere.

The probability of a gene crossing a hybrid zone depends on its degree of linkage to the genes involved in establishing the zone itself, i.e. those involved in reproductive isolation (Barton 1983; Barton and Jones 1983). The mitochondrial genome will inevitably be unlinked to such genes and therefore has the potential for transmission through genetic barriers to hybridization. The theoretical basis of this process has been explored in some detail by Takahata and colleagues (Takahata and Slatkin 1984; Takahata and

Palumbi 1985). There are indeed cases in organisms where hybridization is known to have occurred in the past, such as the amphibian genera *Rana, Hyla,* and *Xenopus,* where incongruities in the distribution of mtDNA restriction maps have been observed and could readily be explained by this process. However, even in amphibians there are examples where the barriers to gene flow across a hybrid zone appear to extend to the mtDNA (Szymura *et al.* 1985).

Since these cases were first reported, several other similar observations have been made. In almost all cases the introgression explanation has been offered, although in some there is little evidence to support it over the retention of an ancestral type. This was particularly true for the two cases found in *Drosophila* (Solignac *et al.* 1986; Satta *et al.* 1988; Powell 1983), where there is not much in the way of independent evidence to support the claims of introgression, except that it is known that the species involved can be hybridized in the laboratory. On the other hand, in *Clethrionomys,* evidence from geographic localization of the invading mtDNA type is again available to support the introgression hypothesis (Tegelstrom 1987). These and other studies have recently been reviewed by Harrison (1989).

Restriction analysis of mtDNA is clearly extremely useful for exploring population structure and will undoubtedly continue to be used on a widening array of species. The haploid, maternal mode of inheritance can result in a four-fold reduction in effective population size relative to nuclear genes (Birky *et al.* 1983). In combination with the relatively high rate of evolution, this makes it a very sensitive marker for subdivision because of the greater sensitivity to the sampling events inherent in the subdivision of populations. Some of the ways in which this may be exploited in future have been indicated by DeSalle and colleagues (DeSalle *et al.* 1987; DeSalle and Templeton 1988; DeSalle *et al.* 1985*a,b*) in studies of Hawaiian *Drosophila* populations. Analysis of populations of the very closely related species *D. sylvestris* and *D. heteroneura* mtDNAs indicate that a significant amount of the observed diversity was due to population structuring (DeSalle *et al.* 1985*a,b*). However, in *D. mercatorum,* quantitative analysis of mtDNA restriction map frequencies was necessary to detect subdivision of the populations (DeSalle *et al.* 1987).

In parallel, recent studies on *D. simulans* and *D. subobscura* (Baba-Aissa *et al.* 1988; Latorre *et al.* 1986) have shed light on the genetic correlates of large-scale population expansion. In both species the colonization of the Americas from the Old World has been associated with a significant loss of mtDNA diversity. Thus we may look forward to the use of mtDNA restriction site analysis in studies not just of the present but also of the past history of populations.

In part because of the generally high level of variability, restriction enzyme studies of mtDNA are able to detect a large number of polymorphic sites. Nucleotide sequencing has not been widely adopted because, given sufficient variability, restriction enzyme analysis is able to handle larger samples from more populations. Where sequencing has been used, in the D loop region of human mtDNA (Aquadro and Greenberg 1983), it was suggested that restriction enzymes could underestimate variation at some sites because of mutational convergence. Aquadro and Greenberg (1983) observed such a strong bias towards transition substitutions (A⇌G and C⇌T), as opposed to transversions (A,G⇌C,T), that the reappearance of an identical restriction site following the loss could not be regarded as wholly unlikely. Thus, in the EcoRI recognition sequence GAATTC, if an adenine mutated to guanine: GAGTTC and the site was lost, it would then be about 10 times as likely to return to adenine than to become either thymine or cytosine. The resulting convergence in nucleotide sequence could produce an underestimate of the extent of nucleotide divergence between two sequences (see Templeton 1983a). This underestimate is a function of the evolutionary distance between the sequences and so would not be expected seriously to affect analyses of intraspecific divergence.

One way in which mtDNA investigations have altered the approaches taken by population geneticists has already been mentioned. Relationships between populations are frequently analysed by taking a phylogenetic approach. This uses the whole array (or haplotype) of restriction sites as characters rather than the classical genetic distance based on gene frequencies at individual sites. Such a phylogenetic method represented a new departure for population genetics. We shall see later that it has proved useful in the analysis of variation in nuclear genes. In Mendelian population genetics, even under the infinite allele model (Kimura and Crow 1964), the possibility of such an approach had not been envisaged.

2.2   Restriction map analysis of multigene families

Restriction enzyme studies have made a fundamental contribution towards understanding the evolution of multigene families. Hood and colleagues first suggested in 1975 that the members of families such as the immunoglobulin genes might not evolve as independent genetic loci (Hood et al. 1975), but the generality of the proposal was established later by restriction map work. It was shown for many different gene families, including satellite sequences, ribosomal genes (rDNA), histone, and other protein-coding genes in Drosophila, primates, mice, and sea urchins, (Arnheim 1983; Jeffreys 1979; Leigh Brown and Ish-Horowicz 1981; Coen et al. 1982; Strachen et al. 1982; Zimmer et al. 1980), that the restriction maps of individual genes were more similar than was expected. In many cases this was done by comparing the restriction maps of family members from closely related species. If the multigene families pre-dated the separation of the two species, and sequences within each shared species-specific restriction sites then the members of these gene families could not have evolved independently. The process of sequence homogenization (concerted evolution, Zimmer *et al.* 1980), is known to proceed via unequal crossing-over, gene conversion or both. The population genetics of sequence evolution within such systems has been extensively studied by Ohta and others (reviewed by Ohta 1985, 1988; see also references in Nagylaki 1988). The evolution of repetitive DNA has been reviewed by Miklos (1985).

For the last few years there have been relatively fewer contributions from restriction site studies and greater emphasis on DNA sequences (Strachan *et al.* 1985; Smithies and Powers 1986). The technical reason for the shift is that molecular cloning purifies a single family member whose exact sequence can then be obtained. Genomic restriction digests do not provide general information on sequence similarities within multigene families, because on autoradiographs where 100 or more copies of a sequence share one particular size of fragment, the 101st which happens to be a different size will not be detected because of the extreme over-exposure of the main band. Indeed, there could be 50 more copies, each with a different fragment size, and none would be detected. Only a few enzymes give informative and interpretable patterns. Their use has therefore tended to result in an over-simplification of the sequence relationships among family members. That these are in fact very complex is indicated by examining the data in Fig. 3 of Strachan *et al.* (1985) (Fig. 1). These sequences were used as illustrations of concerted evolution: at the sites denoted by #, sequences from the two species are fixed for different nucleotides. The restriction pattern for an enzyme which included any of these sites would be very different. However, the total sequence divergence down each branch has only been 1.5 per cent (Fig. 1, caption). Perhaps in consequence there has been a tendency towards over-simplification in discussions of gene family evolution (Dover 1982). On the other hand, cloning and sequencing studies have of necessity been restricted to a very small fraction of the total number of members of each gene family, and results based on sampling a few out of several hundred sequences are subject to large errors.

Most recently, carefully designed restriction site studies have been favoured once again. Seperack *et al.* (1988) have estimated the disequilibrium between two polymorphic sites within the human rDNA repeat. Each haplotype class could be distinguished on an autoradiograph on the basis of band size and the numbers of genes within each class were quantified by densitometry. Significant departures from random association were found among the rDNA families from each of 25 individuals from two populations. The rDNA in humans is located on five different chromosomes, which segregate randomly at meiosis. It can be concluded that the non-random associations are *within* the subfamilies on each chromosome, in which the effect of concerted evolution has already been shown to be most pronounced (Krystal *et al.* 1981). Thus, every generation inherits distinguishable blocks of rDNA repeats which give rise to a significant level of inter-individual heterogeneity over the long term. This study constituted a direct test of the

```
              110        120        130        140        150 #      160        170        180        190  #     200
COM    AAAGTGAAAG GGGTTGTTAG TATTGGTTGT AAGGAGTACA AAATGGTACT NCTTTTGCTC TCTGACCATT TTTAGTCAAG TTATAGCCAA AANAGCCAAT
mau 1  ---------- ---------- ---------- ---------- ---------- T--------- ---------- ---------- ---------- --C-------
    2  -----T---- -C-------- -----T--AC --------T- ---C---*-- T--------- GG------C- A----C---- ---------- --A------C
    3  ---------- ---------- ---------- ---------* ---------- T--------- ---------- ---------- ---------- --C-------
    4  ---------- ---------- ---------- ---------- ---------- T--------- ---------- ---------- ---------- --C-------
    5  ---------- ---------- ---------- ---------- ---------- T--------- ---------- ---------- ---------- --C-------
    6  ---------- --A------- ---------- ---------- ---------- T--------- ---------- ---------- ---------- --C-------
sim 1  ---------- ---------- ------C--- --T-G----- ---------- C--------- ---------- ---------- ---------- --A-------
    2  ---------- ---------- ---------- ---------- ---------C C--------- ---------- ---------- ---------- --A-------
    3  -----T---- --A------- ---------- --------A- ---------- C------G-- ---------- ---------- ---------- --AC------
    4  ---------- ---------- ---------- ---------- ---------- C--------- ---------- ---------- ---------- --A*------
    5  ---------- ---------- ---------- ---------- ---------- C--------- ---------- -------G-- ---------- --AT------
    6                        ---------- ---------- T--------- CA-------- ---------- ---------- ---------- --A--T----
    7                        -----C---- ---*-A---- ---------- C--------- ---------- ---------C ---------- --AC------
    8  ---------* ---------- ---------- ---------- ---------- C--------- ---------- ---------- ---------- --A-------
    9  ---------- ---------- ---------- ---------- ---------- C--------- ---------- ---------- ---------- --A-------

              210        220  #     230        240        250        260  #     270        280        290        300
CON    TTGAAATTTC ATTTTTTTGC CAAAANTATT TTCCCAGATT TTTTGTGAAA AAAATATTTG GTNTTTTGAT CAAAACATTC GAAATAATTA CCCAAATATG
mau 1  ---------- ---------- -----G---- ---------- ---------- ---------- --A------- ---------- ---------- ----------
    2  ------A--G -G-------- -----A--G- ---G--C-*- -------*-- --------C- --A------- ---------G --------G- ----------
    3  ---------- ---------- -----G---- ---------- -------*-- ---------- --A------- ---------- ---------- ----------
    4  ---------- ---------- -----G---- ---------- ---------- ---------- --A------- ---------- ---------- ----------
    5  ---------- ---------- -----G---- ---------- ---------- ---------- --A------- ---------- ---------- ----------
    6  ---------- ---------- -----G---- ---------- ---------- ---------- --A------- ---------- ---------- ----------
sim 1  ---------- ---------- -----A---- ---------- ---------- ---------- --T------- G---*----- ---------- ----------
    2  ---------- ---------- -----A---- ---------- ---------- ---------- --T----*-A ---------- ---------- ---------*
    3  -----C---- -------*-- -----A---- ---------- ---------- ---------- --A------- -----T---- ---------- ----------
    4  ---------- ---------- -----A---- --------*- ---------- ---------- --A------- ----------
    5  ---------- -G-------- -----A---- ---------- ---------- -----T---- --T------- ----*----- ---------- ---
    6  ---------- -*-------A -----A---- A-------*- ---------- ---------- --T------- ---------- ---------* ----------
    7  ---------- ---------- -----A---- ---------- ---------- ---------- --T------- -********* ---------- ----------
    8  ---------- ---------- -----A---- ----------
    9  ---------- ---------G ----CA---- ---------- ---------- ---------- --T------- ---------- ---------- ----------
```

Fig. 1. DNA sequence variation among cloned copies of the '360' sequence from two *Drosophila* species (bases 100–300). (From Strachan *et al.* 1985.) Restriction enzyme studies could seriously overestimate sequence divergence between these two species as only variation that affected a substantial number of copies of the gene family could be detected. The mean nucleotide diversity within each species is 3.4 per cent for *D. mauritiana* and 4.53 per cent for *D. simulans*. The observed divergence between copies of different species is 5.44 per cent. The divergence down each lineage from the polymorphic common ancestor is estimated to be only 1.5 per cent $= (0.0544 - (0.034 + 0.0453)/2)$ (Nei and Tajima 1981). mau = *Drosophila mauritiana*, sim = *Drosophila simulans*, # = nucleotide sites showing concerted evolution.

'molecular drive' hypothesis of Dover (1982), which predicts the contrary: that such heterogeneity would be insignificant in comparison to the total sequence variability in the population. Instead, the result favoured the view that individual heterogeneity remains important in the evolution of rDNA and probably other multigene families (Ohta 1985).

The population genetic analysis of multigene families was not developed as a response to DNA-based results but to those of an earlier generation of molecular genetics based on protein sequencing (Ohta 1978, 1980). However, evidence that unequal crossing-over and gene conversion could be responsible for concerted evolution came from restriction mapping and DNA sequencing studies (Zimmer *et al.* 1980; Slightom *et al.* 1981; Mellor *et al.* 1983). In particular, the significance of gene conversion in the evolution of multigene families has received a considerable amount of attention because of the potential importance of a conflict between selection at the molecular level through biased gene conversion (Dover 1982) and selection at the level of the organism (Walsh 1985, 1987*b*; Nagylaki 1988 and references therein). Equally, although gene duplication itself had been known for a long time, studies on the interaction between the selective and mutational forces that act on duplicated genes (e.g. Stephan 1986; Walsh 1987*a*,*b*) were stimulated by the direct observations which had been made on variation and evolution in such sequences using molecular techniques.

2.3 Evolution of single copy nuclear DNA sequences

2.3.1 *Interspecific divergence at restriction sites*
Early studies of the restriction maps of protein coding genes and flanking regions showed a considerable level of variation within species and of divergence between related species (Jeffreys 1979). Following the development of simple statistical procedures for estimating sequence divergence from restriction map data (Nei and Li 1979; Engels 1981; Ewens *et al.* 1981) some general features became apparent. Nucleotide divergence between *Drosophila melanogaster* and its sibling species *D. mauritiana* and *D. simulans* was estimated to be about 3 per cent in three regions of the genome: the two loci, 87A7 and 87C1, encoding the major heat shock protein (hsp 70) and the *Adh* locus (Leigh Brown and Ish-Horowicz 1981; Langley *et al.* 1982). In primates, however, nucleotide divergence in the β-globin gene cluster was much less. The estimated divergence between human and baboon was 4 per cent, whereas that between human and gorilla was only 0.9 per cent (Barrie *et al.* 1981). These results showed that the morphologically almost identical *Drosophila* species were almost as strongly diverged as human and baboon and much more so than human and gorilla.

The use of restriction analysis of nuclear DNA as a phylogenetic tool has not been extensive. In general, other techniques, such as DNA cloning and sequencing, can provide appreciably more information for a similar amount of work. There has been considerable discussion of the methodology involved in phylogenetic analysis of restriction site data (Templeton 1983*a*; Nei and Tajima 1985; Saitou and Nei 1986; Smouse and Li 1987). The data have been considered in at least two ways: either converted into a distance matrix of nucleotide divergence estimates, or the sites can be considered as characters. In the first case it has generally been assumed that both restriction sites and nucleotide substitutions are distributed randomly along the DNA sequence and the rate of substitution is constant in all lineages. The UPGMA method has often been used to generate a phylogenetic (strictly, phenetic) tree from such a distance matrix. If restriction sites are considered as characters, a maximum parsimony or compatibility method is often used to construct a cladogram. For many data sets, such as those discussed above, the two approaches will not be in conflict. However, if the divergence between the species studied is much greater, or if there is serious violation of the underlying assumptions, some data sets may not give concordant results. This problem will naturally be most serious when the evolutionary distances between the branch-points are small.

The classic example of a data set which has proved recalcitrant is that collected by Ferris *et al.* (1981) and Brown *et al.* (1982) on restriction site divergence among the apes. Despite extensive later analysis (Adams and Rothman 1982; Templeton 1983*a,b*; Nei and Tajima 1985; Saitou and Nei 1986; Smouse and Li 1987) there is little confidence that the correct order of branching has been obtained. The extent of divergence is small between extant lineages and the ancestral groups were undoubtedly polymorphic, and it has been argued that mtDNA restriction data may never be able to provide a robust branching order because of the large number of nucleotides which will need to be screened (Saitou and Nei 1986). The consequence has been a considerable refinement in the methodological aspects without an appreciable advance in our understanding of the true mtDNA phylogeny. The separate issue of whether the mtDNA phylogeny would necessarily reflect the species phylogeny has been discussed in an earlier section.

2.3.2 *Interspecific divergence in restriction map length*

The results of the studies on the β-globin, heat shock and *Adh* genes were in substantial agreement on a second issue. In each case the overall organization of the region of DNA under study, with respect to coding sequences and to the distance between conserved restriction sites (referred to here as 'map length'), remained the same in all these comparisons, apart from the occasional loss of restriction sites. Only in comparisons with much greater divergence in restriction sites was significant change generally seen in the organization of the gene clusters or in the distance between flanking restriction sites. A notable exception to this generalization is the *D. melanogaster* 87C heat shock locus (Leigh Brown and Ish-Horowicz 1981) where centromeric repetitive DNA elements have inserted as a long tandem array between two coding sequences which are arranged as a simple inverted repeat in all its close relatives. A similar dramatic alteration in restriction map length has been reported by Cheng and Hardison (1988) for the rabbit α-globin cluster. Here, a block containing several coding elements extending over nearly 20 kb appears to have been duplicated in one haplotype. Also Mills *et al.* (1986) have described alterations in the distance between duplicated copies of the alcohol dehydrogenase coding sequence among cactophilic drosophilids related to *D. mojavensis*. However, these observations appear to be exceptional.

The evolution of the alcohol dehydrogenase locus has been investigated in Hawaiian picture-wing drosophilids by Bishop and Hunt (1988). The level of nucleotide divergence among the five species studied, estimated from the restriction site distances, is similar to that of the *D. melanogaster/simulans/ mauritiana* group. There was no evidence for substantial change in restriction map length over nearly 20 kb. An even greater level of conservation of map length was described by Payant *et al.* (1988) at the *Amylase* gene regions in the *D. melanogaster* species subgroup. The two coding sequences at the *Amy* locus form an inverted repeat, as with the heat shock genes. In both cases there is evidence for concerted evolution between the repeated copies. However, the distance between two conserved Sal I restriction sites, one in each *Amy* coding sequence, is 6.4 kb in all eight species studied, which included the relatively distantly related *D. erecta* and *D. teisseri*.

There are two possible explanations for the conservation of restriction map length; either (1) restriction map length is less liable to mutation than the nucleotide sequence of non-coding DNA, or (2) there are stronger selective constraints on the length of DNA between genes than on its actual base sequence. As we will see below, the former point is far from being the case, so the indication is that changes in the length of non-coding DNA are subject to natural selection, although how they are expressed phenotypically is unclear.

2.4 Intraspecific variation in single copy genes

Results from studies of intraspecific variation in restriction map do not show the concordance between mammals and *Drosophila* observed in interspecific comparisons of restriction map length. I shall review this area in greater detail, and examine two particular topics: first some of the results that have been obtained on heterozygosity and linkage disequilibrium by analysis of restriction map variation, and secondly, the frequency and nature of restriction map variation due to DNA insertion and deletion.

2.4.1 *Intraspecific divergence at restriction sites—data from* Drosophila

Many recent studies of intraspecific variation at restriction sites have adopted the approach of Kreitman and Aguade (1986*a*), because it is possible to increase the sensitivity of the survey if enzymes which recognize 4 base-pair sites are used. Many more nucleotide sites can be screened with a given probe, and also much smaller deletion and insertion events can be scored. The distribution of fragment sizes produced by cleavage with such enzymes inevitably has a much smaller mean. With an expected fragment size of the order of 250–350 base-pairs quite different gel techniques are required. The technical advances made by Kreitman and Aguade (1986*a,b*) lay in the combination of gel systems developed for DNA sequencing with sensitive hybridization protocols. In consequence they were able to resolve polymorphisms in fragments as small as 60 base-pairs reliably and score a total of 954 nucleotide 'site-equivalents' in a 6.9 kb region of the *Adh* locus in 80 lines of *D. melanogaster* from two different populations. Kreitman (1987) estimates that as many as 25 per cent of all nucleotide substitutions may be detected with this technique.

Several surveys of intraspecific restriction site variation in *D. melanogaster* have now been completed. The use of the 4-base cutter technique has greatly extended the early work on the *Adh* and 87A heat shock loci. I have summarized the results for seven regions of the *D. melanogaster* genome, distributed in each of the major chromosomes (Table 3*a*). All have been screened in a sample of about 30 or more chromosomes and the length of DNA screened, the number of restriction sites detected, and the preparation of polymorphic sites are listed. The level of variation for each locus (heterozygosity per nucleotide) is indicated by the values for two estimators. The values obtained with these two estimators rarely differ greatly. The variability covers a range from 0.002/0.003 for the *achaete-scute*, heat shock and *rosy* loci to 0.008/0.009 for *Amylase* and *white*. On average we could expect between 1/100 and 1/300 nucleotides to be heterozygous in this species.

The study of intraspecific restriction site variation has been extended to two other species: *D. pseudoobscura* for the *Adh* locus and *D. simulans* for the *rosy* locus (Schaeffer *et al.* 1987; Aquadro *et al.* 1988). It is interesting to note that in both cases a significantly higher heterozygosity per nucleotide was found in these other species than in *D. melanogaster*, four-fold higher for *Adh* in *D. pseudoobscura* and six-fold higher for *rosy* in *D. simulans* (Table 3*b*). These observations present an intriguing contrast with protein polymorphism in these species where the heterozygosity of *D. melanogaster* is greater than that of *D. simulans* (Ohnishi *et al.* 1982; Choudhary and Singh 1987). Both these observations could be accommodated under the view that *D. melanogaster* has a lower effective population size (Aquadro *et al.* 1988). Then the heterozygosity due to completely neutral mutations would be lower—as for restriction sites, while a higher level of variation due to nearly

Table 3a
Restriction site variation in D. melanogaster *populations*

Locus	white[1]	Notch[2]	A–SC[3]	Adh[4-6]	Amy[7]	87A[8] Heat shock	rosy[9]
				II		III	
DNA screened (kb)	45 9.6*	60	120	13 6.9*,5	13	25	40
Restriction sites	56 271*	58	69	30 238*	26	25	41
Polymorphic sites	19 42*	15	6	8 25*	6	2	7
θ†	0.008 0.004*	0.007	0.0021	0.006	0.006		
π‡	0.009 0.004*	0.005	0.0023		0.008	0.0024	0.003
Sample size	64	37	49	49	85	29	60

Table 3b
Restriction site variation in other species

Locus	D. pseudoobscura Adh	D. simulans rosy
DNA screened (kb)	32	40
Restriction sites	43	56
Polymorphic sites	27	28
θ†	0.026	
π‡		0.019
Sample size	20	30

*Screened with 4-base cutting enzymes. †θ estimated by the procedure of Hudson (1982).
‡π estimated following Nei and Tajima (1981).
[1]Myashita and Langley (1988). [2]Schaeffer *et al.* (1988). [3]Beech and Leigh Brown (1989).
[4]Aquadro *et al.* (1986). [5]Kreitman and Aguade (1986a,b). [6]Aguade (1988). [7]Langley *et al.*
(1988). [8]Leigh Brown (1983). [9]Aquadro *et al.* (1988).

neutral mutations (Ohta 1972, 1973) would be expected. Allozyme variation in *D. melanogaster* could fall into this latter category.

The analysis of more than one species also permits detailed conclusions to be drawn on the forces controlling the evolution of a particular gene or DNA region. In an earlier volume of this series Kreitman (Kreitman 1987) discussed the use of sequence data at the *Adh* locus of two *Drosophila* species in a test of the neutral theory of Kimura (1983) (Hudson *et al.* 1987). The availability of sufficiently unambiguous molecular, in that case sequence, data has stimulated this rigorous approach, although unfortunately only the neutral theory provides sufficiently explicit predictions in terms of allele frequencies to be tested in this way.

2.4.2   *Restriction site variation in a virus: HIV–1*

A recent example of how restriction site studies can provide data from any organism that can then be used in a population genetic analysis was given by the publication of an analysis of variation in HIV–1 molecules cloned in bacteriophage lambda (Saag *et al.* 1988). From nine to 17 HIV genomes were cloned from each of three viral cultures, two of which, WMF1 and WMF3, were established from samples taken from the same patient at different times. A total of 47 restriction sites detected in the 9.8 kb genome with 11 6-cutter restriction enzymes were scored in all 62 clones obtained from these three samples (the full data set included another five unrelated HIV–1 clones in which 22 more restriction sites were detected). Of the 47 sites, only 15 did not show any variation. This remarkable level of variability is characteristic of HIV–1 (see e.g. Hahn *et al.* 1986) in which sequence evolution occurs even more rapidly than in other retroviruses (Leigh Brown and Monaghan 1988).

Considering the three samples separately, the level of within-sample nucleotide diversity is much lower (Table 4). Thus each sample consists of clones that are significantly more closely related to each other than to independent samples. Secondly, the nucleotide diversity estimates ('heterozygosity' is not an appropriate term) show that within each sample a high level of variability can be detected but that the level is similar (0.11–0.12) in each group (Table 4). Thirdly, the two samples from viral cultures established from the same patient show a significant increase in diversity, approaching half that observed between the samples from different patients, although the two samples were isolated only 16 months apart.

The population genetics methodology which has been developed for molecular data can be readily applied to situations such as this which are unfamiliar to many evolutionary biologists. The issues are in essence the same: what forces are controlling the frequencies of various genotypes? In the particular case of HIV the mutation rate is elevated many orders of magnitude above that of even mtDNA, but the pattern of within and between sample diversity is similar to that found in many studies on more or less isolated populations of other organisms. I believe we may expect the extension of population genetics into the viral kingdom to continue in view of the increasing significance of viral diseases for health care in the developed world. The urgency associated with studies of nucleotide sequence variation of HIV–1 is likely to make it a special case of a general trend.

Table 4
Nucleotide diversity within three HIV-1 isolates from two patients

Virus isolate	RJS4	WMF1	WMF3	WMF1 + WMF3
No. of sites screened	32	35	33	38
No. of variable sites	12	13	12	21
No. of HIV clones in sample	27	17	18	35
Nucleotide diversity*	0.120	0.116	0.107	0.210
Nucleotide distance from RJS4*	–	0.403	0.441	(ND)

*From Leigh Brown and Balfe (1988); data from Saag *et al.* (1988).

2.4.3 *Linkage disequilibrium between restriction sites in* Drosophila

Although a newly arising mutation is found in only one combination of linked polymorphic variants, recombination causes these associations to decay with time. However, natural or artificial selection can generate non-random associations of existing alleles at different loci (gametic disequilibrium) when there is epistasis. In some cases this can lead to tight linkage of the loci involved into so-called 'supergenes' (Hedrick *et al.* 1978). Such associations can also be generated in small populations by sampling effects (Hill and Robertson 1966, 1968). The demonstration of abundant polymorphism at the protein level using biochemical techniques led to a huge increase in the number of studies on disequilibrium in nature, motivated by the expectations of the Franklin–Lewontin model of selection (Franklin and Lewontin 1970) and urged on by Lewontin's eloquent exposition (Lewontin 1974). Significant disequilibria were not usually observed, but it was acknowledged that allozyme data proved not to be ideal for the purpose (reviewed by Langley 1977). Polymorphic restriction sites are generally much more closely linked, can be unambiguously scored and represent a precise genetic locus, so in recent years restriction site data have come to be used extensively in studies of disequilibrium in nature.

In genetical terms, restriction maps represent a very fine-scale analysis and restriction sites near the same gene or gene cluster will usually be very closely linked. It was therefore not surprising that early studies detected linkage disequilibrium among polymorphic restriction sites. Such disequilibria were found in both *Drosophila* and mammalian genes (Langley *et al.* 1982; Antonarakis *et al.* 1982; Leigh Brown 1983). There has recently been a significant re-evaluation of the level of disequilibrium within restriction maps of *Drosophila* genes and I will discuss these data first, returning to some of the mammalian results later.

Although there is little differentiation between loci with respect to nucleotide variability, there is much wider divergence in the extent of linkage disequilibrium among polymorphic restriction sites at each locus. In this respect the loci used in the first surveys of restriction site variation—the *Adh* and 87A heat shock loci—turned out not to be representative. At both, significant associations were found between restriction map variants 5 or even 10 kb apart (Leigh Brown 1983; Aquadro *et al.* 1986; Cross and Birley 1986). As more loci have been studied, the occurrence of such disequilibria has declined. Thus at the *Notch* and *Amylase* loci, only a few pairwise comparisons are significant. In the latter case these are between restriction sites and the allozyme phenotype, whereas in the former the number was very low in relation to the number of tests performed.

This last point has become a problem as the scale of the surveys has increased. The *rosy* and *white* loci also failed to reveal extensive disequilibria. In the study of the *white* region, the number of pairwise comparisons is so large (over 3500) that 35 positive results could be expected at a probability of $P < 0.01$ by chance. Thus the authors have to use extra information in the form of the distance between the sites in order to establish which tests might indeed be significant and which are likely to be type I errors. There was in fact pronounced clustering of the disequilibrium values in the transcription unit, a non-random pattern which suggests that these are real associations.

At most loci summarized in Table 5, the frequency of observing significant disequilibria between sites separated by 5 kb or more is low. Indeed, at the *white* locus, variants only 100–200 base-pairs apart are frequently in linkage equilibrium. However the *achaete-scute* region is very different in this respect. At this locus significant disequilibria are common and have been observed between sites separated by distances of 65 and 80 kb (Beech and Leigh Brown 1989; Macpherson, Leigh Brown, and Weir, in preparation). Recently, the analysis at *achaete-scute* was extended to test for three-way disequilibria. Associations between variants at three loci are more difficult to detect in samples of moderate size (Brown 1977), but significant associations which spanned a total distance of 70–80 kb were found. It is known that recombina-

Table 5
Disequilibrium analyses in D. melanogaster *populations*

	white[1]	Notch[2]	A–SC[3]	Adh[4-6]	Amy[7]	87A[8] Heat shock	rosy[9]
No. of tests	3570	55	27	66	6	3	3
No. significant	103	6	17	16	2*	3	0
Propn. tests significant	0.03	0.11	0.63	0.24	0.33	1	0

*Both of these tests involved the *Amy* electrophoretic phenotype.
[1]Myashita and Langley (1988). [2]Schaeffer *et al.* (1988). [3]Beech and Leigh Brown (1989). [4]Aquadro *et al.* (1986). [5]Kreitman and Aguade (1986*a,b*). [6]Aguade (1988). [7]Langley *et al.* (1988). [8]Leigh Brown (1983). [9]Aquadro *et al.* (1988).

tion is suppressed in this region of the X chromosome (Dubinin *et al.* 1937), and so some increase in the frequency of significant disequilibria might be expected. However, the observation that variants approaching 100 kb apart are more strongly associated than sites about 100 base-pairs apart at *white* was unexpected. These results suggest that this region will evolve in a much more complex manner than other parts of the genome, with blocks of DNA which include several transcribed sequences, being transmitted in an almost clonal fashion. It will be interesting to discover how the complex has changed over evolutionary time, and whether specific haplotypes have differentiated significantly in sequence.

2.4.4 *Linkage disequilibrium in mammalian genes*

Analysis of polymorphic restriction sites in the human β-globin gene cluster quickly revealed a striking pattern of disequilibrium (Antonarakis *et al.* 1982; Kazazian *et al.* 1983; Chakravarti *et al.* 1984). Eight restriction sites spread over the entire cluster are polymorphic in all human races. Pairwise disequilibrium values between many of these polymorphic sites have been shown to be significant in several populations. However, in this gene cluster the probability of a disequilibrium value being significant is not simply a function of the molecular distance between the restriction sites. The restriction sites fell into two groups, one spanning 35 kb between the ε- and ψβ-genes and another spanning 19 kb to the 3′ side of the β-globin gene. Within each group, significant disequilibria are found frequently and in more than one population regardless of the distance between the restriction sites. Between the two groups, disequilibria are rarely significant even though the distance between the margins of the groups is only about 10 kb (Antonarakis *et al.* 1982; Chakravarti *et al.* 1984). Although recombination has been detected in the region around the human β-globin gene (Gerhard *et al.* 1984), there is unfortunately little knowledge of the genetic map distances involved.

The major histocompatibility complex (MHC) of the mouse, the H–2 gene cluster on chromosome 17 is a gene complex the genetics of which is much better understand than that of the β-globin region. There are five genes in this cluster which are highly polymorphic. Two of these five genes, H–2K and H–2D, encode transplantation antigens and are known as class I genes, while three are class II genes (I-A$_\alpha$, I-A$_\beta$ and I-E$_\beta$) involved in the function of T helper cells. There are many other genes (over 50 coding sequences) which show lower levels of variability within the H–2 complex. At the DNA level, the entire gene cluster may span some 2500 kb or more and a genetic distance of about 2 centimorgans (reviewed by Klein 1987).

Considerable divergence in restriction map has been found between inbred strains of mice and the variable sites have been used to map the position of cross-over events detected immunologically (Steinmetz *et al.* 1982). A high proportion of the cross-overs have apparently occurred within short regions of DNA (Steinmetz 1986).

The genetic map distance between the I-A$_\beta$ and I-E$_\alpha$ genes is 0.1 per cent (Snell *et al.* 1976), and they are 85 kb apart on the molecular map (Steinmetz *et al.* 1984). However, restriction mapping has shown that all the recombination events on which this genetic distance was based have occurred within 10 kb of the E gene, and probably within the 1.7 kb long large intron in this gene (Steinmetz *et al.* 1982). In that particular interval, cross-overs involving three haplotypes, *k, b*, and *s* have all been observed. There are other 'hot spots' in the complex which seem to be specific to certain haplotypes. In one such case, the subspecies *M. musculus castaneus* has generated several recombinants between the K and I-A$_\beta$ genes, five of which have been localized to within 3.5 kb. As all five involved only the same *M.m. castaneus* haplotype, there is clearly some sequence-specificity in the 'hot spot' effect.

Recombination 'hot spots' are clearly important for determining the type of recombinants which arise at greatest frequencies. As 'hot spots' appear to be sequence-specific, they will differ in their location in related populations. This will have complex implications for our expectations of disequilibrium in nature. One interesting feature has arisen from comparisons among the H–2 restriction maps of three unrelated mouse strains in the light of these recombination 'hot spots'. Cosmid clones covering almost the entire I region have been obtained from three strains of mice, BALB/c (H–2^d haplotype), AKR (H–2^k haplotype), and B10.WR7 (H–2^{WR7}) (Steinmetz *et al.* 1984). Restriction maps for eight restriction enzymes have been obtained for the entire I gene region. Up to 100 sites have been scored in each strain and a high level of polymorphism is present. However, different segments within the region differ significantly in their heterozygosity. In Table 6 I give the heterozygosity per nucleotide calculated by Hudson's (1982) method from the data given by Steinmetz *et al.* (1984). The heterozygosity is presented separately for the I–A region of the cluster and the I–E region, with the divisions being made at the I–E$_\beta$ recombination 'hot spot'. The three-fold difference in variability between these two regions is contributed equally by all pairwise comparisons. The higher variability of the I–A region does not, however, seem to extend to the extreme end of the cloned region, available only in B10^{WR7} and Balb/c. This confirms the conclusions of Steinmetz *et al.* (1984) who used a different measure of variability.

I have discussed elsewhere how such a discordance in the distribution of polymorphic sites could arise as a consequence of selection and linkage (Leigh Brown 1989). In brief, it is probably reasonable to assume that the high level of serologically detected variability at the I–A$_\alpha$ and I–E$_\beta$ genes is maintained by some form of balancing selection. This assumption would extend the conclusions of Hedrick and Thomson (1983) and Klitz *et al.* (1986) on the human class I genes to the murine class II genes but is probably realistic. Restriction sites proximal to a recombination 'hot spot' will be very tightly linked to the coding sequences in that region. From Steinmetz' results (Steinmetz 1986), crossing the 'hot spot' would reduce the level of linkage

Table 6

Nucleotide heterozygosity in the I region of the mouse H–2 complex. (Based on data from Steinmetz et al *1984)*

Gene region*	Coordinates	Strains compared	Heterozygosity Mean	SD†
1. I–A$_\beta$, I–A$_\alpha$	44–120	BALBc/B10.WR7	0.069	0.017
		BALBc/AKR	0.066	0.014
		AKR/B10.WR7	0.048	0.012
		BALBc/AKR/B10.WR7	0.0714	0.015
2. I–E$_{\beta2}$, I–E$_\alpha$	120–270	BALBc/B10.WR7	0.027	0.007
		BALBc/AKR	0.013	0.005
		AKR/B10.WR7	0.028	0.007
		BALBc/AKR/B10.WR7	0.022	0.006

*Division between the two regions made at the recombination 'hot spot' in the I–E$_\beta$ large intron.

†Estimated by the method of Hudson (1982).

considerably. It is well known that selection can alter the level of heterozygosity at linked neutral loci. In the case of balancing selection this is known as 'associative over-dominance' (Hill and Robertson 1968; Sved 1968; Ohta and Kimura 1969, 1970) or 'pseudo-selection' (Thomson 1977) and has the result that the heterozygosity at very closely linked loci will be elevated. As the linkage between the selected and neutral loci declines, the effect decreases rapidly. This explanation for the unusual observations on the H–2 gene cluster has the attraction that it can be distinguished from the alternative proposed by Steinmetz (1986): that the tracts of high and low variability have arisen as a by product of gene conversion events. Although it is known that gene conversion involving distant donor sequences is an important source of new variation in coding regions (Mellor *et al.* 1983), Steinmetz *et al.* (1986) would require that sequence homology extends for a wide region outside the coding sequences themselves.

2.4.5 *Polymorphism in restriction map due to DNA insertion*

A considerable body of work has found that genetic variation due to the insertion of large DNA segments is relatively common at many gene regions in *D. melanogaster* (Langley *et al.* 1982; Leigh Brown 1983; Aquadro *et al.* 1986; Langley and Aquadro 1987; Beech and Leigh Brown 1989; Schaeffer *et al.* 1988). The origin of these insertions was shown by Leigh Brown (1983) and Aquadro *et al.* (1986) to be the insertion of transposable genetic elements. Leigh Brown (1983) identified a P element in one chromosome and an mdg4 (gypsy) element in another which had inserted close to the sequences encoding the major heat shock protein at the 87A7 heat shock locus. Aquadro *et al.* (1986) similarly described insertions of *copia*, B104,

2161, and 101F near the *Adh* coding sequence. The observations of such large DNA insertions in close proximity to coding sequences was in itself remarkable as the chromosomes into which the insertions had occurred usually had normal viability and fertility. Indeed their abundance suggests that, on average, chromosomes in nature probably carry a large DNA insertion every few 100 kb. A second unusual feature of this type of variation was that in general, particular insertions were rare, often unique in the sample. Both these features of DNA insertions could be interpreted in the light of surveys of variation in the chromosomal location of transposable elements in this species, the results of which have been reviewed by Charlesworth (1989) and Leigh Brown (1989). Although there are many questions still outstanding, the observations suggest that insertions of transposable elements have mild deleterious effects which act in opposition to the increase in number of insertions that would result from the replicative transposition of such elements (Charlesworth and Charlesworth 1983; Golding *et al.* 1986; Montgomery *et al.* 1987; Langley *et al.* 1988).

If so, this would explain why changes in restriction map length are only rarely observed between closely related species (see above), despite the intraspecific variability. However, studies on *Adh* in *D. pseudoobscura* and the *rosy* gene region in *D. simulans* suggest that variation due to DNA insertion is very much less abundant in these species (Schaeffer *et al.* 1988; Aquadro *et al.* 1988). This was particularly striking because in both cases heterozygosity due to nucleotide substitution was considerably higher than in *D. melanogaster* (see above) so that the difference could not be attributed to problems with sampling. As mentioned earlier, the difference in heterozygosity due to nucleotide substitution has reasonably been interpreted as indicating that *D. melanogaster* has a significantly lower effective population size than these species (Aquadro *et al.* 1988). This could be due to a recent expansion from a much lower species population size, an explanation which makes sense for a species whose distribution is almost inextricably tied to human activities.

There is an interesting corollary to this suggestion. It is well known that the number of slightly deleterious mutations which will behave as if they are neutral increases as the effective population size declines (Kimura 1968; Ohta 1972, 1973). This applies whether the deleterious effect is recessive, as suggested for transposable element DNA insertions by B. Charlesworth and D. Charlesworth (1983) and Beech and Leigh Brown (1989), or partially dominant as proposed by Montgomery *et al.* (1987) and Langley *et al.* (1988). A corresponding increase in heterozygosity is expected due to previously deleterious mutations now behaving as if they are neutral. In view of the evidence that such insertions are deleterious, the suggestion on independent data that *D. melanogaster* has a lower effective population size might thus provide an attractive explanation for the presence of a class of variability which is clearly absent in otherwise apparently similar *Drosophila* species. Whether differences in effective population size could be sufficient to explain the difference in total amount of mobile repetitive DNA between species is another but intriguing issue. A considerable effort has been made towards understanding the population genetic forces which regulate transposable element copy number (B. Charlesworth and D. Charlesworth 1983; Langley *et al.* 1983; Kaplan and Brookfield 1983; Charlesworth 1985; Charlesworth and Langley 1986; Montgomery *et al.* 1987; Langley *et al.* 1988; Charlesworth 1989). Nevertheless, we do not have a complete explanation for the distribution of insertions observed in *D. melanogaster* and would therefore appear to be still some way from directly addressing the difference in *D. simulans* in anything other than the most general terms.

3. CONCLUSIONS—RESTRICTION ENZYMES, A TOOL FOR THE FUTURE?

3.1 Technical aspects

Although restriction enzymes were first being employed some 10 years ago as tools in the analysis of genetic variability, there has been a significant lag before they became more generally adopted. They have undoubtedly opened up new possibilities in population genetic analysis. It is likely that a wider range of groups will continue to apply the methodology to problems of population structure and, perhaps to a lesser extent, to surveys of genetic variability in a widening range of species. Whether restriction map analysis will continue to be used as the main method for surveying genetic variability is more questionable. The method by which they may be supplanted is already in use in the laboratories of some population geneticists: the polymerase chain reaction, introduced by Saiki *et al.* (1985) has now been simplified and automated following the incorporation of the heat-stable DNA polymerase from *Thermophilus aquaticus*. DNA prepared by this method, which has been shown to be able to generate significant quantities of DNA from a single DNA molecule obtained from an isolated sperm cell (Li *et al.* 1988), can be sequenced directly (McMahon *et al.* 1987). For surveys of nucleotide variability, or studies of disequilibrium, it is likely to be the method of the early 1990s.

3.2 General

The subjects covered in this review are not linked by a conceptual theme but by a technical one. There is, obviously, a range of evolutionary questions for which a molecular approach is not possible, or may not seem to be appropriate. However, the influence of developments in molecular genetics should not be underestimated at a time when, for example, the US Congress has adopted, in principle, a proposal to sequence the entire human genome, or, more relevantly, when many of the key genes which control the establishment of the embryonic body plan of *Drosophila* have been identified (Ingham 1988). This convergence of genetics with developmental biology is now being extended to take in quantitative characters (Michelmore and Shaw 1988; Patterson *et al.* 1988; Shrimpton and Robertson 1988). We may reasonably expect that in the future a population biologist can request, or even buy, cloned probes for genes which will contribute a defined proportion of the additive genetic variance in almost any trait. At such a time, it will be possible to take a molecular approach to almost any evolutionary question, given that there is an underlying genetic component.

This 'brave new world' (taken in the sense of William Shakespeare or Aldous Huxley according to your presentiments) is foreshadowed here, not by the details of results obtained from restriction enzyme studies but rather by the way in which, in several different areas, the novel data have thrown up unforeseen questions. In consequence, the theory of population genetics now extends to a much wider variety of situations and has in many cases come closer to evolutionary reality than was possible in the past because of the lack of appropriate data. The process must surely accelerate.

ACKNOWLEDGEMENTS

I am grateful to Linda Partridge, Jim Bull, and Paul Harvey for their comments.

REFERENCES

Adams, J. and Rothman, E. D. (1982). Estimation of phylogenetic relationships from DNA restriction patterns and selection of endonuclease cleavage sites. *Proc. Natl Acad. Sci. USA,* **79,** 3560–4.

Aguade, M. (1988). Restriction map variation at the *Adh* locus of *Drosophila melanogaster*, in inverted and noninverted chromosomes *Genetics* **119,** 135–40.

Antonarakis, S. E., Boehm, C. D., Giardina, P. J., and Kazazian, H. H. (1982). Non-random association of polymorphic restriction sites in the β-globin cluster. *Pro. Natl Acad. Sci. USA,* **79,** 137–41.

Aquadro, C. F. and Greenberg, B. D. (1983). Human mitochondrial DNA variation and evolution: analysis of nucleotide sequences from seven individuals. *Genetics* **103,** 287–312.

——, Deese, S. F., Bland, M. M., Langley, C. H., and Laurie-Ahlberg, C. C. (1986). Molecular population genetics of the alcohol dehydrogenase gene region of *Drosophila melanogaster*. *Genetics* **114,** 1165–90.

——, Lado, K. M., and Noon, W. A. (1988). The *rosy* region of *Drosophila melanogaster* and *Drosophila simulans*. I. Contrasting levels of naturally occurring restriction map variation and divergence. *Genetics* **119,** 875–88.

Arnheim, N. (1983). Concerted evolution of multigene families. In *Evolution of genes and proteins* (ed. M. Nei and R. K. Koehn), pp. 38–61. Sinauer Associates, Sunderland, Massachusetts.

Ashley, M. and Wills, C. (1987). Analysis of mitochondrial DNA polymorphisms among Channel Island deer mice. *Evolution* **41**, 854–63.

Avise, J. C. (1986). Mitochondrial DNA and the evolutionary genetics of higher animals. In *The evolution of DNA sequences.* (ed. B. C. Clarke, A. Robertson and A. J. Jeffreys) *Phil. Trans. Roy. Soc. Lond.* **B312**, 325–42.

——, Lansman, R. A., and Shade, R. O. (1979). The use of restriction endonucleaes to measure mitochondrial DNA sequence relatedness in natural populations. I. Population structure and evolution in the genus Peromyscus. *Genetics* **92**, 279–95.

——, Shapira, J. F., Daniel, S. W., Aquadro, C. F., and Lansman, R. A. (1983). Mitochondrial DNA differentiation during the speciation process in Peromyscus. *Molec. Biol. Evol.* **1**, 38–56.

——, Neigel, J. E., and Arnold, J. (1984). Demographic influences on mitochondrial DNA lineage survivorship in animal populations. *J. Mol. Evol.* **20**, 99–105.

——, *et al.* (1987). Intraspecific phylogeography: the mitochondrial DNA bridge between population genetics and systematics. *Ann. Rev. Ecol. Syst.* **18**, 481–522.

Baba-Aissa, F., Solignac, M., Dennebourg, N., and David, J. R. (1988). Mitochondrial DNA variability in *Drosophila simulans*: quasi-absence of polymorphism within each of the three cytoplasmic races. *Heredity* **61**, 419–26.

Barrie, P. A., Jeffreys, A. J., and Scott, A. F. (1981). Evolution of the β-globin gene cluster in man and the primates. *J. Mol. Biol.* **149**, 319–36.

Barton, N. H. (1983). Multilocus clines. *Evolution,* **37**, 454–71.

—— and Jones, J. S. (1983). Mitochondrial DNA: new clues about evolution. *Nature,* **306**, 317–18.

Beech, R. N. and Leigh Brown, A. J. (1989). Insertion-deletion variation at the yellow, achaete-scute region in two natural populations of *Drosophila melanogaster*. *Gen. Res. Camb.* **53**, 7–15.

Berger, S. L. and Kimmel, A. R. (1987). Guide to molecular cloning techniques. In *Methods in enzymology*, Vol. 152. Academic Press, Orlando.

Bickle, T. A. (1987). DNA restriction and modification systems. In Escherichia coli *and* Salmonella typhimurium: *Cellular and molecular biology*, (ed. F. C. Neidhart). American Society for Microbiology, Washington DC.

Birky, C. W. Jr, Maruyama, T., and Fuerst, P. (1983). An approach to population and evolutionary genetic theory for genes in mitochondria and chloroplasts, and some results. *Genetics* **103**, 513–27.

Bishop, J. G. III and Hunt, J. A. (1988). DNA divergence in and around the alochol dehydrogenase locus in five closely related species of Hawaiian *Drosophila*. *Mol. Biol. Evol.* **5**, 415–31.

Brown, A. H. D. (1977). Sample sizes required to detect linkage disequilibrium between two or three loci. *Theoret. Popul. Biol.* **8**, 184–201.

Brown, W. M., Prager, E. M., Wang, A., and Wilson, A. C. (1982). Mitochondrial DNA sequences of primates: tempo and mode of evolution. *J. Mol. Evol.* **18**, 225–39.

Chakravarti, A. *et al.* (1984). Non-uniform recombination within the human β-globin gene cluster. *Amer. J. Hum. Genet.* **36**, 1239–58.

Charlesworth, B. (1985). The population genetics of transposable elements. In

Population genetics and molecular evolution (ed. T. Ohta and K. Aoki), pp. 213–32. Springer, Berlin.

—— (1989). The population biology of transposable elements. In *Evolution and animal breeding* (ed. W. G. Hill and T. F. C. Mackay), pp. 53–9. C.A.B. International, Wallingford.

—— and Charlesworth, D. (1983). The population dynamics of transposable elements. *Genet. Res. Camb.* **42**, 1–27.

—— and Langley, C. H. (1986). The evolution of self-regulated transposition of transposable elements. *Genetics* **112**, 359–83.

Cheng, J. F., and Hardison, R. C. (1988). The rabbit α-like globin gene cluster is polymorphic both in the sires of BamHI fragments and in the number of duplicated sets of genes. *Mol. Biol. Evol.* **5**, 486–98.

Choudhary, M. and Singh, R. S. (1987). A comprehensive survey of genic variation in natural populations of *Drosophila melanogaster*. III. Variations in genetic structure and their causes between *Drosophila melanogaster* and its sibling species *Drosophila simulans*. *Genetics* **117**, 697–710.

Church, G. M. and Gilbert, W. (1984). Genomic sequencing. *Proc. Natl Acad. Aci. USA*, **81**, 1991–5.

Coen, E. S., Strachan, T., and Dover, G. A. (1982). The dynamics of concerted evolution of rDNA and histone gene families of *melanogaster* species subgroup of *Drosophila*. *J. Mol. Biol.* **158**, 17–35.

Cohen, S., Chang, A., Boyer, H., and Helling, R. (1973). Construction of biologically functional bacterial plasmids *in vitro*. *Proc. Natl Acad. Sci. USA*, **70**, 3240–4.

Cross, S. R. H. and Birley, A. J. (1986). Restriction endonuclease map variation in populations of *Drosophila melanogaster*. *Biochem. Genet.* **24**, 415–33.

DeSalle, R. and Templeton, A. R. (1988). Founder effects and the rate of mitochondrial DNA evolution in Hawaiian Drosophila. *Evolution* **42**, 1076–84.

——, Giddings, L. V., and Templeton, A. R. (1985*a*). Mitochondrial DNA variability in natural populations of Hawaiian *Drosophila*. I. Methods and levels of variability in natural populations of *D. sylvestris* and *D. heteroneura*. *Heredity* **56**, 75–85.

——, Giddings, L. V., and Kaneshiro, K. Y. (1985*b*). Mitochondrial DNA variability in natural populations of Hawaiian *Drosophila*. II. Genetic and phylogenetic relationships of natural populations of *D. sylvestris* and *D. heteroneura*. *Heredity* **56**, 87–96.

——, Templeton, A., Mori, I., Pletscher, S., and Johnston, J. S. (1987). Temporal and spatial heterogeneity of mtDNA polymorphisms in natural populations of *Drosophila mercatorum*. *Genetics* **116**, 215–23.

Dover, G. (1982). Molecular drive, a cohesive mode of species evolution. *Nature* **299**, 11–117.

Dubinin, N. P., Sokolov, N. N., and Tiniakov, G. G. (1937). Crossing over between the genes *yellow, achaete* and *scute*. *Dros. Inf. Serv.* **8**, 76.

Engels, W. R. (1981). Estimating genetic divergence and genetic variability with restriction endonucleases. *Proc. Natl Acad. Sci. USA* **78**, 6329–33.

——, Spielman, R. S., and Harris, H. (1981). Estimation of genetic variation at the DNA level from restriction endonuclease data. *Pro. Natl Acad. Sci. USA* **788**, 3748–50.

Feinberg, A. O. and Vogelstein, B. (1984). A technique for radiolabelling DNA restriction endonuclease fragments to high specific activity. *Anal. Biochem.* **132,** 6–13.

Ferris, S. D., Wilson, A. C., and Brown, W. M. (1981). Evolutionary tree for apes and humans based on cleavage maps of mitochondrial DNA. *Proc. Natl Acad. Sci. USA* **78,** 2432–6.

—— *et al.* (1983). Flow of mitochondrial DNA across a species boundary. *Proc. Natl Acad. Sci. USA* **80,** 2290–4.

Franklin, I. and Lewontin, R. C. (1970). Is the gene the unit of selection? *Genetics* **65,** 707–34.

Gerhard, D., Kidd, K. K., Kidd, J. R., Egeland, J. A., and Housman, D. E. (1984). Identification of a recent recombination event within the human β-globin cluster. *Proc. Natl Acad. Sci. USA* **81,** 7875–9.

Golding, G. B., Aquadro, C. F., and Langley, C. H. (1986). Sequence evolution within populations under multiple types of mutation. *Proc. Natl Acad. Sci. USA* **83,** 427–31.

Hahn, B. H. *et al.* (1986). Genetic variation in HTLV-III/LAV over time in patients with AIDS or at risk from AIDS. *Science* **232,** 1548–53.

Haldane, J. B. S. (1927). A mathematical theory of natural and artificial selection, Part V: Selection and mutation. *Proc. Camb. Philos. Soc.* **23,** 838–44.

Harrison, R. G. (1989). Animal mitochondrial DNA as a genetic marker in population and evolutionary biology. *Trends Ecol. Evol.* **4,** 6–11.

Hedrick, P. W. and Thomson, G. (1983). Evidence for balancing selection at HLA. *Genetics* **104,** 449–56.

——, Jain, S., and Holden, L. (1978). Multilocus systems in evolution. *Evolutionary biology*, vol. 11, (ed. M. K. Hecht, W. C. Steere and B. Wallace) pp. 101–84. Plenum, New York.

Hill, W. G. and Robertson, A. (1966). The effect of linkage on the limits to artificial selection. *Genet. Res. Camb.* **8,** 269–94.

—— and —— (1968). Linkage disequilibrium in finite populations. *Theor. Appl. Genet.* **38,** 226–31.

Hood, L., Campbell, J. H., and Elgin, S. C. R. (1975). The organisation, expression and evolution of antibody genes and other multigene families. *Ann. Rev. Genet.* **9,** 305–52.

Hudson, R. R. (1982). Estimating genetic variability with restriction endonucleases. *Genetics* **100,** 711–19.

——, Kreitman, M., and Aguade, M. (1987). A test of neutral molecular evolution based on nucleotide data. *Genetics* **116,** 153–9.

Ingham, P. W. (1988). The molecular genetics of embryonic pattern formation in *Drosophila*. *Nature* **335,** 25–34.

Jeffreys, A. J. (1979). DNA sequence variants in the G_γ, A_γ, δ-and β-globin genes of man. *Cell* **18,** 1–10.

—— and Flavell, R. A. (1977). A physical map of the DNA regions flanking the rabbit β-globin gene. *Cell* **12,** 429–39.

Kaplan, N. L. and Brookfield, J. F. Y. (1983). Transposable elements in Mendelian populations. III. Statistical results. *Genetics* **104,** 485–95.

Kazazian, H. H., Chakravarti, A., Orkin, S. H., and Antonarakis, S. E. (1983). DNA polymorphisms in the human β-globin cluster. In *Evolution of genes and proteins* (ed. M. Nei and R. K. Koehn), pp. 137–46. Sinauer Associates, Sunderland, Massachusetts.

Klitz, W., Thomson, G., and Baur, M. P. (1986). Contrasting evolutionary histories among tightly linked HLA loci. *Amer. J. Hum. Genet.* **39**, 340–9.

Kimura, M. (1968). Evolutionary rate at the molecular level. *Nature* **217**, 624–6.

—— (1983). *The neutral theory of molecular evolution.* Cambridge University Press, Cambridge.

—— and Crow, J. F. (1964). The number of alleles that can be maintained in a finite population. *Genetics* **49**, 725–38.

Kreitman, M. (1983). Nucleotide polymorphism at the alcohol dehydrogenase locus of *Drosophila melanogaster. Nature* **304**, 412–17.

—— (1987). Molecular population genetics. *Oxford surveys in evolutionary biology*, vol. 4 (ed. P. Harvey and L. Partridge), pp. 38–60. Oxford University Press, Oxford.

—— and Aguade, M. (1986a). Genetic uniformity in two populations of *Drosophila melanogaster* revealed by filter-hybridisation of four nucleotide recognising restriction enzyme digests. *Proc. Natl Acad. Sci. USA* **83**, 3562–6.

—— and —— (1986b). Excess polymorphism at the *Adh* locus in *Drosophila melanogaster. Genetics* **114**, 93–110.

Krystal, M., D'Eustachio, P., Ruddle, F. H., and Arnheim, N. (1981). Human nucleolus organisers on non-homologous chromosomes can share the same ribosomal gene variants. *Proc. Natl Acad. Sci. USA* **78**, 5744–8.

Langley, C. H. (1977). Non-random associations between allozymes in natural populations of *Drosophila melanogaster.* In *Measuring selection in natural populations* (ed. F. B. Christiansen and T. M. Fenchel), pp. 265–73. Springer, Berlin.

—— and Aquadro, C. F. (1987). Restriction map variation in natural populations of *Drosophila melanogaster: white*-locus region. *Mol. Biol. Evol.* **4**, 651–630.

——, Montgomery, E., and Quattlebaum, W. F. (1982). Restriction map variation in the *Adh* region of *Drosophila. Proc. Natl Acad. Sci. USA* **79**, 5631–5.

——, Brookfield, J. F. Y., and Kaplan, N. L. (1983). Transposable elments in Mendelian populations. I. A theory. *Genetics* **104**, 457–72.

—— et al. (1988). Naturally occurring restriction map variation of the *Amy* region of *Drosophila melanogaster. Genetics* **119**, 619–29.

Latorre, A., Moya, A., and Ayala, F. J. (1986). Evolution of mitochondrial DNA in *Drosophila subobscura. Proc. Natl Acad. Sci. USA* **83**, 8649–53.

Leigh Brown, A. J. (1983). Variation of the 87A heat shock locus in *Drosophila melanogaster. Proc. Natl Acad. Sci. USA* **80**, 5350–4.

—— (1989). Restriction map analysis in population genetics. In *Evolution and animal breeding*, (ed. W. G. Hill and T. F. C. Mackay) pp. 39–44. C.A.B. International, Wallingford.

—— and Balfe, P. (1988). Relationships among isolates of HIV. *Nature* **335**, 675.

—— and Ish-Horowicz, D. (1981). Evolution of the 87A and 87C heat shock loci in *Drosophila. Nature* **290**, 677–82.

—— and Monaghan, P. (1988). Evolution of the structural proteins of human immunodeficiency virus: selective constraints on nucleotide substitution. *AIDS. Res. Hum. Retroviruses* **4**, 399–407.

Lewontin, R. (1974). *The genetic basis of evolution change.* Columbia University Press, New York.

Li, H. *et al.* (1988). Amplification and analysis of DNA sequences in single human sperm and diploid cells. *Nature* **335**, 414–7.

McMahon, G., Davis, E., and Wogan, G. N. (1987). Characterisation of c-Ki-*ras* oncogene alleles by direct sequencing of enzymatically amplified DNA from carcinogen-induced tumours. *Proc. Natl Acad. Sci. USA* **84**, 4974–8.

MacNeil, D. and Strobeck, C. (1987). Evolutionary relationships among colonies of Columbian Ground Squirrels as shown by mitochondrial DNA. *Evolution* **41**, 873–81.

Maxam, A. M. and Gilbert, W. (1977). A new method for sequencing DNA. *Proc. Natl Acad. Sci. USA* **74**, 560–4.

Mellor, A. L., Weiss, E. H., Ramachandran, K., and Flavell, R. A. (1983). A potential donor gene for the *bm1* gene conversion event in the c57BL mouse. *Nature* **306**, 792–5.

Michelmore, R. W. and Shaw, D. V. (1988). Character dissection. *Nature* **335**, 672–3.

Miklos, G. G. L. (1985). Localised highly repetitive DNA sequences in vertebrate and invertebrate genomes. In *Molecular evolutionary genetics* (ed. R. J. MacIntyre), pp. 131–240. Plenum Press, New York.

Mills, L. E. P., Batterham, P., Alegre, J., Starmer, W. T., and Sullivan, D. T. (1986). Molecular genetic characterisation of a locus that contains duplicate Adh genes in *Drosophila mojavensis* and related species. *Genetics* **112**, 295–310.

Modrich, P. and Roberts, R. J. (1982). Type-II restriction and modification enzymes. In *Nucleases* (ed. S. M. Linn and R. J. Roberts). Cold Spring Harbor, New York.

Montgomery, E. A., Charlesworth, B., and Langley, C. H. (1987). A test for the role of natural selection in the stabilisation of transposable element copy number in a population of *Drosophila melanogaster*. *Genet. Res., Camb.* **49**, 31–41.

Moritz, C., Dowling, T. E., and Brown, W. M. (1987). Evolution of animal mitochondrial DNA: relevance for population biology and systematics. *Ann. Rev. Ecol. Syst.* **18**, 269–92.

Moriwaki, K. *et al.* (1984). Implications of the genetic divergence between European wild mice with Robertsonian translocations from the viewpoint of mitochondrial DNA. *Genet. Res., Camb.* **43**, 277–87.

Nagylaki, T. (1988). Gene conversion, linkage and the evolution of multigene families. *Genetics* **120**, 291–301.

Nei, M. and Li, W.-H. (1979). Mathematical model for studying genetic variation in terms of restriction endonucleases. *Proc. Natl Acad. Sci. USA* **76**, 5269–73.

—— and Tajima, F. (1981). DNA polymorphism detectable by restriction endonucleases. *Genetics* **97**, 145–63.

—— and —— (1983). Maximum likelihood estimation of the number of nucleotide substitutions from restriction sites data. *Genetics* **105**, 207–17.

—— and —— (1985). Evolutionary change of restriction cleavage sites and phylogenetic inference for man and apes. *Molec. Biol. Evol.* **2**, 189–205.

Neigel, J. E. and Avise, J. C. (1986). Phylogenetic relationships of mitochondrial DNA under various demographic models of speciation. In *Evolutionary processes and theory* (ed. E. Nevo and S. Karlin), pp. 515–34. Academic Press, New York.

Nelson, K., Baker, R. J., and Honeycutt, R. L. (1987). Mitochondrial DNA and protein differentiation between hybridising cytotypes of the white-footed mouse, *Peromysucus leucopus*. *Evolution* **41**, 864–72.

Ohnishi, S., Leigh Brown, A. J., Voelker, R. A., and Langley, C. H. (1982).

Estimation of genetic variability in natural populations of *Drosophila simulans* by two-dimensional and starch gel electrophoresis. *Genetics* **100**, 127–36.

Ohta, T. (1972). Population size and evolution. *J. Mol. Evol.* **1**, 305–14.

—— (1973). Slightly deleterious mutant substitutions in evolution. *Nature* **246**, 96–8.

—— (1978). Sequence variability of immunoglobulins considered from the standpoint of population genetics. *Proc. Natl Acad. Sci. USA* **75**, 5108–12.

—— (1980). *Evolution and variation of multigene families.* Lecture notes in biomathematics, Vol. 37. Springer, Berlin.

—— (1985). Genetic variation of multigene families. In *Population genetics and molecular evolution* (ed. T. Ohta and K. Aoki), pp. 233–42. Springer, New York.

—— (1988). Multigene and supergene families. In *Oxford surveys in evolutionary biolology*, Vol. 5 (ed. P. Harvey and L. Partridge). Oxford University Press.

—— and Kimura, M. (1969). Linkage disequilibrium at steady state determined by random genetic drift and recurrent mutation. *Genetics* **63**, 229–38.

—— and —— (1970). Development of associative overdominance through linkage disequilibrium in finite populations. *Genet. Res., Camb.* **16**, 165–77.

Palmer, J. D. (1985). Evolution of chloroplast and mitochondrial DNA in plants and algae. In *Molecular evolutionary genetics* (ed. R. J. MacIntyre), pp. 131–240. Plenum Press, New York.

Patterson, A. H. *et al.* (1988). Resolution of quantitative traits into Mendelian factors by using a complete linkage map of restriction fragment length polymorphisms. *Nature* **335**, 721–6.

Payant, V. *et al.* (1988). Evolutionary conservation of the chromosomal configuration and regulation of amylase genes among eight species of the *Drosophila melanogaster* species subgroup. *Mol. Biol. Evol.* **5**, 560–7.

Powell, J. R. (1983). Interspecific cytoplasmic gene flow in the absence of nuclear gene flow: evidence from *Drosophila*. *Proc. Natl Acad. Sci. USA* **80**, 492–5.

Ratner, L. *et al.* (1985). Complete nucleotide sequence of the AIDS virus, HTLV-III. *Nature* **313**, 277–83.

Riesner, D. and Gross, H. H. (1985). Viroids. *Ann. Rev. Biochem.* **54**, 531–64.

Rigby, P. W. J., Dieckman, M., Rhodes, C., and Berg, P. (1977). Labelling deoxyribonucleic acid to high specific activity *in vitro* by nick translation with DNA polymerase I. *J. Mol. Biol.* **113**, 237–51.

Saag, M. S. *et al.* (1988). Extensive variation of human immunodeficiency virus type-1 *in vivo*. *Nature* **334**, 440–4.

Saitou, N. and Nei, M. (1986). The number of nucleotides required to determine the branching order of three species, with special reference to the human-chimpanzee-gorilla divergence. *J. Mol. Evol.* **23**, 189–204.

Sanchez-Pescador *et al.* (1985). Nucleotide sequence and expression of an AIDS-associated retrovirus (ARV-2). *Science* **227**, 484–92.

Sanger, F., Nicklen, S., and Coulson, A. R. (1977). DNA sequencing with chain-terminating inhibitors. *Proc. Natl Acad. Sci. USA* **74**, 5463–8.

Satta, Y. *et al.* (1988). Dubious maternal inheritance of mitochondrial DNA in *D. simulans* and evolution of *D. mauritiana*. *Genet. Res., Camb.* **52**, 1–6.

Saiki, R. K. *et al.* (1985). Enzymatic amplification of globin genomic sequences and restriction site analysis for diagnosis of sickle cell anaemia. *Science* **230**, 1350–4.

Schaeffer, S. W., Aquadro, C. F., and Langley, C. H. (1988). Restriction-map variation in the *notch* region of *Drosophila melanogaster*. *Mol. Biol. Evol.* **5**, 30–40.

Seperack, P., Slatkin, M., and Arnheim, N. (1988). Linkage disequilibrium in human ribosomal genes: implications for multigene family evolution. *Genetics* 119, 943–9.

Shah, D. M. and Langley, C. H. (1979). Inter- and intra-specific variation in restriction maps of Drosophila mitochondrial DNAs. *Nature* 281, 696–9.

Shields, G. F. and Wilson, A. C. (1987). Subspecies of the Canada Goose (*Branta canadensis*) have distinct mitochondrial DNAs. *Evolution* 41, 662–6.

Shrimpton, A. E. and Robertson, A. (1988). The isolation of polygenic factors controlling bristle score in *Drosophila melanogaster*. I. Allocation of third chromosome sternopleural bristle effects to chromosome sections. *Genetics* 118, 437–43.

Slighton, J. L., Blechl, A. E., and Smithies, O. (1981). Human fetal $^G\gamma$- and $^A\gamma$-genes: complete nucleotide sequences suggest that DNA can be exchanged between these duplicated genes. *Cell* 21, 627–38.

Smithies, O. and Powers, P. A. (1986). Gene conversions and their relation to homologous chromosome pairing. In *The evolution of DNA sequences* (ed. B. C. Clarke, A. Robertson, and A. J. Jeffreys). *Phil. Trans. Roy. Soc. Lond.* B312, 227–42.

Smouse, P. E. and Li, W. H. (1987). Likelihood analysis of mitochondrial restriction-cleavage patterns for the human-chimpanzee-gorilla trichotomy. *Evolution* 41, 1162–76.

Snell, G. D., Dausset, J., and Nathenson, S. (1976). *Histocompatability*. Academic Press, New York.

Solignac, M., Monnerot M., and Monoulou, J.-C. (1986). Mitochondrial DNA evolution in the *melanogaster* species subgroup of *Drosophila*. *J. Mol. Evol.* 23, 31–40.

Southern, E. M. (1975). Detection of specific sequences among DNA fragments separated by el electrophoresis. *J. Mol. Biol.* 98, 503–17.

Spolsky, C. and Uzzell, T. (1984). Natural interspecies transfer of mitochondrial DNA in amphibians. *Proc. Natl Acad. Sci. USA* 81, 5802–5.

Steinmetz, M. (1986). Polymorphism and recombinational hot spots in the murine MHC. *Curr. Topics Microbiol. Immunol.* 127, 279–84.

—— et al. (1982). A molecular map of the immune response region from the major histocompatibility complex of the mouse. *Nature* 300, 35–42.

—— et al. (1984). Tracts of high or low sequence divergence in the mouse major histocompatibility complex. *EMBO J.* 3, 2995–3003.

Stephan, W. (1986). Recombination and the evolution of satellite DNA. *Genet. Res., Camb.* 47, 167–74.

Strachan, T., Coen, E. S., Webb, D. A., and Dover, G. A. (1982). Concerted rearrangements in the evolution of abundant DNA families in the *melanogaster* species subgroup. *J. Mol. Biol.* 158, 37–54.

——, Webb, D., and Dover, G. A. (1985). Transition stages of molecular drive in multiple-copy DNA familes in Drosophila. *EMBO J.* 4, 1701–8.

Strobeck, C. and Morgan, K. (1978). The effect of intragenic recombination on the number of alleles in a population. *Genetics* 88, 829–44.

Sved, J. A. (1968). The stability of linked systems of loci with a small population size. *Genetics* 59, 543–63.

Szymura, J. M., Spolsky, C., and T. Uzzell. (1985). Concordant change in mitochondrial and nuclear genes in a hybrid zone between two frog species (genus *Bombina*). *Experientia* 41, 1469–70; 26, 375–400.

Takahata, N. and Palumbi, S. R. (1985). Extranuclear differentiation and gene flow in the finite island model. *Genetics* **109**, 441–57.

—— and Slatkin, M. (1984). Mitochondrial gene flow. *Proc. Natl Acad. Sci. USA* **81**, 1764–7.

Tegelstrom, H. (1987). Transfer of mitochondrial DNA from the northern red-backed vole (*Clethrionomys rutilus*) to the bank vole (*C. glareolus*). *J. Mol. Evol.* **24**, 218–27.

Templeton, A. R. (1983*a*). Convergent evolution and non-parametric inferences from restriction site data and DNA sequences. In *Statistical analysis of DNA sequence data* (ed. B. S. Weir). Marcel Dekker, New York.

—— (1983*b*). Phylogenetic inference from restriction endonuclease cleavage site maps with particular reference to the evolution of humans and the apes. *Evolution* **37**, 221–44.

Thomson, G. (1977). The effect of a selected locus on linked neutral loci. *Genetics* **85**, 753–88.

Wahl, G. M., Stern, M., and Stark, G. R. (1979). Efficient transfer of large DNA fragments from agarose gels onto diazobenzyloxymethyl paper and rapid hybrisiation using dextran sulfate. *Proc. Natl Acad. Sci. USA* **76**, 3683–7.

Wain-Hobson, S., Sonigo, P., Danos, O., Cole, D., and Alizon, M. (1985). Nucleotide sequence of the AIDS virus, LAV. *Cell* **40**, 9–17.

Walsh, J. B. (1985). Interaction of selection and biased gene conversion in a multigene family. *Proc. Natl Acad. Sci. USA* **82**, 153–7.

—— (1987*a*). Sequence-dependent gene conversion; can duplicated genes diverge fast enough to escape conversion? *Genetics* **117**, 543–57.

—— (1987*b*). Persistence of tandem arrays: implications for satellite and simple-sequence DNA's. *Genetics* **115**, 553–67.

Weissman, C. (1989). Sheep disease in human clothing. *Nature* **338**, 298–9.

Wilson, A. C. *et al.* (1985). Mitochondrial DNA and two perspectives on evolutionary genetics. *Biol. J. Linn. Soc.* **26**, 375–400.

Yuan, R. (1981). Structure and mechanism of multifunctional restriction endonucleases. *Ann. Rev. Biochem.* **50**, 285–315.

Zimmer, E. A., Martin, S. L., Beverley, S. M., Kan, Y. W., and Wilson, A. C. (1980). Rapid duplication and loss of genes coding for the α-chain of hemoglobin. *Proc. Natl Acad. Sci. USA* **77**, 2158–62.

Index

OXFORD SURVEYS IN EVOLUTIONARY BIOLOGY

Volume 1: 1984

Contents

OXFORD SURVEYS IN EVOLUTIONARY BIOLOGY
Volume 2: 1985

Contents

OXFORD SURVEYS IN EVOLUTIONARY BIOLOGY

Volume 3: 1986

Contents

OXFORD SURVEYS IN EVOLUTIONARY BIOLOGY

Volume 4: 1987

Contents

OXFORD SURVEYS IN EVOLUTIONARY BIOLOGY

Volume 5: 1988

Contents

OXFORD SURVEYS IN EVOLUTIONARY BIOLOGY

Volume 5: 1988

Contents

Oxford Surveys in Evolutionary Biology

Oxford Surveys in Evolutionary Biology may be obtained from your local bookseller or on direct subscription through the Journals Subscription Department at Oxford University Press.

Once recorded as a direct subscriber, you will be notified some six to eight weeks prior to the publication of each volume (you may cancel at any time), and **Oxford Surveys in Evolutionary Biology** will be delivered directly to you, immediately on publication.

Simply fill in the reservation form overleaf and return it to the Journals Subscription Department, Pinkhill House, Eynsham, Oxford OX8 1JJ, UK. Back volumes may also be ordered on this form.

We regret, it is not possible to dispatch any volumes until payment has been received.

ORDER FORM

I have purchased Volume 6 of **Oxford Surveys in Evolutionary Biology** and I wish to be recorded as a direct subscriber to future volumes. I understand that I will receive regular notification of the price of future volumes before they are despatched and that I may cancel my subscription at any time.

Name ....

Address ....

....

Postcode .... Country ....

Please supply additional volumes of **Oxford Surveys in Evolutionary Biology** as below.

☐ Volume 1, 1984 £35.00   ☐ Volume 5, 1988 £35.00
☐ Volume 2, 1985 £30.00
☐ Volume 3, 1986 £32.00
☐ Volume 4, 1987 £32.00

☐ I enclose my remittance.
☐ Please debit my Access/American Express/Diner/Visa Account*

* *Delete as appropriate*
Card Number ☐☐☐☐☐☐☐☐☐☐☐☐☐-☐☐☐☐

Expiry Date .... Signature..
If address registered with card company differs from above, please give details.

Please send me details of:
☐ Oxford Reviews of Reproductive Biology
☐ Oxford Surveys in Eukaryotic Genes
☐ Oxford Surveys in Plant Molecular and Cell Biology

Please note that prices given are correct at the time of printing but may be changed at the Publishers discretion and without notice.